The Energy Age

The Energy Age

A guide to the use and abuse of energy
in the world today

Robert Kyriakides

Genersys

Published by Genersys plc

First Published 2006

© Copyright Genersys plc

Genersys plc
37 Queen Anne Street, London W1G 9JB, United Kingdom
www.genersys.com

I have endeavoured to state the facts, as they are known to me
as at 21st March 2006. The views and opinions expressed herein
are mine

British Library Cataloguing-in-Publication Data
A catalogue record for this book is available from the
British Library

ISBN 0-9543232-3-8

Designed by Chris Fayers
Jacket and chart design by Gary Hicks
Edited by Mary Powell

Printed in China

"... but man, proud man,
Drest in a little brief authority,
Most ignorant of what he's most assured,
His glassy essence, like an angry ape,
Plays such fantastic tricks before high heaven
As make the angels weep;"

Isabella, Measure for Measure

Acknowledgements

Many people have helped me both in the preparation of this book but also in the formulation of the ideas that I have set out in it. I must especially thank my colleagues at Genersys who have encouraged and helped me.

I have also, as a result of Genersys' work come into contact with many people at organisations that understand the importance of energy from their own perspectives. In particular I should single out the National Energy Action Charity and the Energy Efficiency Partnership for Homes. They both have powerful lobby voices and are lending them for the good of the environment.

I am much indebted to Mary Powell, who carefully edited this work and provided invaluable advice, without which this book would have been very deficient. The mistakes, however, are my sole responsibility.

Robert Kyriakides

Contents

Preface

Four short years have elapsed since I wrote the first edition of "A Concise Guide to Energy in the United Kingdom". Much has happened in that time. The price of energy has increased significantly. Oil prices have more than doubled in four years and it looks as though they have not yet hit their peaks.

Wind turbines have become much more commonplace and often their installation gives rise to local controversy. Most importantly, India and China are growing economically and vastly, expanding their industrial activities and in consequence using more and more energy not only in manufacturing but also as the living standards of their peoples improve.

Four years on, people are much more familiar with the concepts of renewable energy and global warming. Some install heat pumps and many more take to solar thermal technology. Everyone sees wind farms as they travel the countryside. Everyone suffers energy bills that grow. People hear about, or experience, unusually severe weather.

Governments are moving in the right direction, albeit very slowly. They are being pushed in the right direction by their people.

Energy is a concept that is complex. To understand energy you have to understand physics, economics, human behaviour and commerce. It also helps if you understand construction technologies, sociology, chemistry and biology. Knowing a bit about politics, international affairs and geography is also useful. Perhaps most of all you have to try to understand humankind.

I have included many pictures and diagrams, in the hope that where the words fail the pictures will succeed.

Robert Kyriakides, London, 2006

Chapter 1

The age of energy

We live in the energy age. We are unaware of it, because we take energy for granted but in truth this era is the energy age. Energy is the most critical factor in our lives. The presence or absence of energy causes and will continue to cause nations to become powerful or impoverished. Its price and availability fixes the success or failure of economies and the businesses that trade within and across them. Its absence affects our health welfare and ability to earn our livings. The acquisition of energy has already led to conflict as some nations seek to exploit a position of dominance having large reserves of it. Very rich people and nations use energy in abundance. The poorest people and nations either lack energy or abuse the fuel resources that they have.

Wherever energy is used, its by-products leak into the atmospheric layer that encloses this planet. They affect the air that every living being breathes. By products of energy use leach into the water that we drink and the seas that nourish and sustain life. The waste from energy use becomes an important constituent of the atmosphere, a delicate and complex balance that changes to create unforeseen and unpredictable consequences to our climate and our weather. The use of energy creates the most serious long term threat to human existence, although we do not know for certain whether in this context long term means years, decades, centuries or millennia.

The stone, bronze, iron, industry, science, nuclear and the information technology ages have all described the journey of human history through time; their contributions to our civilisations are still important and will continue to be, but energy has now become the key feature of present and future human progress. That is why we live in the age of energy.

Progress, as my old geography teacher was fond of saying, is simply the putting of one foot in front of another in a particular direction. Progress is not necessarily beneficial. It just happens as a feature of human existence. Progress has moved us into a place where energy is central and critical to all of us. Growth is also a two sided coin. Growth can be positive and healthy and thus can bring undoubted benefits for humankind. Growth can also be cancerous, bringing pain and death to humankind. Progress and growth are made possible by energy; energy is both the catalyst and the foundation.

Despite this, most nations do not understand or plan for energy in a cohesive logical way. Energy ministers, like those of the United Kingdom are usually of junior rank, without seat in the cabinet. They tend to come and go like the newspapers, shuffling off to obscurity or "more important" posts. Decisions about energy are more often avoided than taken, with politicians being reluctant to take unpopular but necessary decisions. That is too hard for them.

In addition economic growth as we now know it is pursued without considering the effect on its consequences. When the fuel for growth is energy and energy is no longer available on the terms we enjoyed during growth, how will we cope?

It seems to me inevitable that the centrality of energy will cause many problems in the future. Nations that have it will want to increasingly keep it for themselves. Those that do not have it will want it. Those that have had it but find that it is now running out will want to secure future supplies. Economic wars and probably armed conflict will occur with increasing frequency over energy and its provider, fuel, as one part of humankind continues to assert itself and its way of life over other parts of humankind.

Therefore it is important to understand what energy is, what its sources are and how much of it we shall continue to need. I have set out in the following chapters our sources of this vital ingredient of life – energy. I describe the traditional sources, and their effects, and some relatively new sources of energy. I have tried, where appropriate, to show energy and the science of energy in its historical context, because that I think leads to a deeper understanding of the issues. In some cases I have given a short account of how certain sources were discovered or used, so that a better understanding of the overall complexity of energy can be reached. It is not, in my view, possible to consider our energy future without understanding how and why we are in the position that we are today.

I have then described how various international groupings and states are dealing with energy in terms of their policy. I have paid particular attention to the United Kingdom and the United States.

I have also described the efforts being made to distribute, control and maintain energy throughout the world. There are many efforts and many good intentions. I also describe the dangers that wide scale energy use brings and may bring. This area often raises the spectre of doom, gloom and calamity with many pressure groups almost vying to see whose prophesies can be the direst. It is important to retain a balanced perspective of the future and a realistic one.

Decisions about energy and in particular energy policy should not be made out of a response to pressure or as a kind of posturing, but should be founded on the best factual knowledge that we have in all the circumstances combined with a single set of ethical principles that should govern us. No one knows when (and possibly if) global warming will irretrievably adversely affect our way of life. No one can know when the fossil fuel resources will be all used up.

But there are many things that we do know and what we do know shows us there are other things that we ought to know. With this knowledge be able to make the many adjustments we need to make when living in the age of energy so that it will not be followed by a long dark age without energy.

The most important thing that we have to do, as a species, as nations, as communities and as individuals is to develop ways of living in the energy age and adjusting to its likely future. A dark age following the energy age may never arise, perhaps by new discoveries, perhaps by prudent human management and excellent governance, but the risk is real and it would be foolish to assume that we do not need to manage the risks of living in the energy age.

Energy is more abused than used. It can be used for good and for evil. Because it is apparently freely available many people use it as though it were an infinite resource. The only factor that limits the indiscriminate use of energy today is its cost. We need to move away from this cost based limitation of energy and towards establishing and applying principles which govern the use of energy.

We must develop personal and national energy policies, and develop them quickly. These must be founded on scientific facts. Our goal must be to attempt to minimise the risks that living in the age of energy brings so that individuals, enterprises and states will behave in ways that minimises these risks.

The risks are twofold; that we run short of energy and that our use of energy harms us irretrievably. I do not think anyone can measure these risks accurately. I think it does not matter whether the risk factors are 99% or 1%. Both percentages are unacceptably high when you consider what we are risking. A future without energy or with energy rationing is just as bleak as a future in a damaged environment.

There is no single solution to the problems of the energy age but there are, I believe, four fundamental principles that should govern our energy age, and are an effective the code of living in the energy age. These principles are all equally important.

The first principle is benign energy first: whenever we can generate energy by benign means, (that is to say without producing carbon dioxide, other greenhouse gas or pollution) we should do so.

The second principle is to conserve energy: whenever we can conserve or save energy or its expenditure, we should do so. At the moment little energy is conserved.

The third principle is that the polluter must pay: whenever energy is used the user should pay not only for the cost of the energy but also the damage that its use does. At the moment society as a whole pays for the damage that energy use does. Without the polluter paying for damage caused by energy people will continue to pollute.

The fourth principle is that there should be no unnecessary use of energy: no one should use products made from energy that is created in a way that causes damage or products that use more energy than is necessary. At the moment few care whether unnecessary energy was expended in the creation or use of products.

The four principles may at first reading appear reasonable and uncontroversial. When we come to apply them we will find that we must change our behaviour in many important ways. They are ways in which no government could countenance without widespread popular support.

Adhering to these four principles require, for example, *several* forms of renewable energy generation by every home. They would mean limiting car engine sizes either by regulation or by effective taxation or a combination of both. They may mean consumers embargoing goods from countries that have produced them in breach of these principles – a massive change of mentality at a time when everyone is pursuing economic growth. They mean changing our agriculture and patterns of consumption. They involve making homes more expensive to build but less expensive to heat and service with energy.

Adopting these principles involves looking at concepts like payback and efficiency with fresh eyes in their true environmental economic and energy contexts. The application of these principles conflicts with much of conventional economics especially that which is espoused by politicians.

Governments in democracies depend upon the votes of the electorate to stay in office. They will not adopt these principles unless there is large scale support for them among the electorate. Is there this level of support in the democracies at present? Energy ministers usually have almost no authority and energy policies are a muddle of wishful thinking and aspirations moulded by pressure from vested interests. We do not simply have to say the "right" thing. We have to do it and do it now.

Chapter 2

Our energy needs

We all need energy to keep warm, to heat and clean our homes, and to warm the water that keeps our clothes and ourselves clean. When we work, we need energy to light our workplaces and power our tools whether they are lathes or computers. We use energy to get us to work and to help us buy our necessities and luxuries. We need energy for our hospitals and for our entertainment. We get most of our energy from gas, oil and coal which either is burnt direct or turned into electricity and petrol.

So, an essential function of all governments is to ensure that there is a supply of plentiful, cheap and accessible energy that is secure and cannot easily be disrupted. This is as important a part of a government's responsibility as defence, law and order, medical services and education. Without energy, our lives would be substantially impoverished and at greater risk.

In modern times, energy policy has been directed to ensuring a diversity of supply so as to prevent one industry or group of interests, like, for example, the British coal industry unions in the early 1970s, or other states and countries, like OPEC behaved in the mid 1970s, from having an undue influence over the United Kingdom by being able to cut off the energy supply. In modern times, securing oil supplies has given the Middle East a political importance that it did not have sixty years ago; many people think that the war in Iraq has had a dimension created by the need of the Western democracies for a secure supply of oil.

Most of our electricity and petrol come from gas and oil that in turn come from huge multinational corporations. Governments must ensure that these corporations do not have excessive power over any state. These corporations are very large and deal in essential products so it is quite easy for them to have, by their sheer size, undue influence without attributing any sinister motives to them. The behaviour of people and corporations and states is largely governed by self interest and energy is so important that it has to come into every equation.

Generally, most countries have tried to cope with these issues by ensuring that they have a diversity of supply. In the United Kingdom the electricity generating industry was originally dominated by coal but then diversified into gas, oil and nuclear power generating stations. Coal gas was replaced by the natural gas that came with the fortunate discovery of oil and gas in commercial quantities in the North Sea.

By the early 1990s the United Kingdom had plenty of relatively cheap energy. A policy of privatisation meant that there were a number of suppliers of energy in each field. Competition meant some control on prices. People used more energy and use of energy grew and grew. As wealth grew, so did travel, with many people holidaying in places where they not only used energy to travel but also increased the energy use in their destinations.

While energy use expanded, there also grew a more and more widespread concern about the effect of its increasing use. More people acquired cars as public transport systems declined or became too expensive or too dangerous; cars were more often than not the cheapest and most convenient way to travel. Cars cause pollution in an obvious way that everyone who stands next to a busy major road can experience. But increased use of energy within residential or business premises also causes pollution, when we burn fuel in order to generate electricity or to heat our homes and work places. This pollution is not as obvious as traffic fumes or smoke billowing from factories but is just as dangerous.

We all consume energy in the form of electricity, gas, oil and other fuel in almost every aspect of our lives from when we wake up in the morning and turn on the light until we go to sleep at night. Even then, energy is being consumed within our homes and outside them for our benefit. There is street lighting, energy used by hospitals and the emergency services and energy used by night workers in baking our bread or making other products that we need.

Modern life inevitably involves many journeys: to work, school and on social occasions. Families no longer invariably live and work close to each other in small communities. While thirty or forty years ago most journeys would have been by public transport, people today value the safety, cleanliness and convenience of driving around in their cars. Other factors of modern life conspire to make the car essential for most families.

Energy plays an important part in the health of people. Cold people die of hypothermia, even in today's Britain. Energy drives life saving hospital equipment. It can prevent people from overheating; in 2003 many deaths were attributed to an unexpectedly hot summer in France. People living without electricity in parts of Africa where the air is clean can develop breathing problems as they burn fuel to provide light in their homes.

Housewives of fifty years ago shopped almost every day for food but working people today are just too busy to be able to shop daily. Instead, they have to buy larger and heavier volumes of shopping. Usually, this means a trip to a supermarket by car. Much of what they buy is needlessly packaged, requiring cars to carry the increased weight and more waste-disposal vehicles to remove it to larger infill sites.

People often expect (or are prepared) to take their children on long journeys to what they perceive to be a better school than the local one. They go on more frequent weekend breaks, fly abroad on holidays and they shop at hours convenient to them at places which are almost inaccessible if they have to use buses and trains. It comes therefore as no surprise that transport is the major energy user.

The amount of energy used for road freight and road passengers per mile of distance travelled has remained fairly constant over the past 25 years, but the huge increase in energy used for transport can be attributed to the large increases in distance now travelled by both passengers and freight and, in particular, the increase in the number of cars on the road. A family car was once a rarity. Nowadays two and three car families are common.

Energy consumption by households for heating water and space, and for cleaning and preparation of food has also gradually increased over the past twenty-five years so that over that whole period it has risen by 21%. Although there are fluctuations from year to year – due to some winters being colder and some summers being hotter – the general trend of domestic consumption of energy is upwards. We expect this trend to continue. The overall growth in energy consumption is thought to be largely due to the increased number of households, rather than individual households using more energy, although we believe that over the next twenty years energy consumption will also increase within each household.

We expect that people will continue to bath and wash more and use more exotic forms of lighting – internally as well as externally for security purposes. More family members will have and use their own individual televisions, music playing equipment, computers, and other electrical equipment. We are no longer one TV families.

While energy consumption has increased significantly over the past 25 years, this increase has taken place against the background of very large energy efficiencies that have disguised the true amount of energy-related activity increases.

During the past 25 years the biggest waste of household energy – poor loft insulation – has been addressed in many buildings. In 1974 only 40% of the housing stock had loft insulation and then the average insulation was only one inch thick. By 1998 almost all housing had loft insulation, most of it being of a reasonable depth to insulate the loft well.

Hot water storage cylinders were also poorly insulated 25 years ago, but modern insulated cylinders have replaced most old cylinders. In addition, many households have installed double-glazing. Cavity wall insulation has become popular for reasons I shall explain later. While this is not suitable for all housing, there is no doubt that once installed, it does bring energy efficiencies, although it is not as effective in saving energy as a well-insulated loft and a well-lagged cylinder.

Although energy saving by insulation has had a discernable effect on energy consumption, overall it has not done more than compensate for the increased use of energy by new appliances and more frequently used machines.

There are also savings in energy consumption that have been made and no doubt will continue to be made by energy efficient white goods. These include fridges, washing machines and tumble dryers where the savings are probably close to optimum. As far as space and water heating are concerned, the development of energy efficient boilers – such as the condensing boiler – can lead to great savings but they are expensive and not suitable for all installations. The government scheme for a £200 refund on the purchase of a condensing boiler was phased out in April 2000 but the government still levy value added tax on condensing boilers and on insulation.

Savings from insulation and energy efficient appliances are modest. We expect more consumption of energy over the next 20 years

High quality insulation, which saves or conserves energy, being installed in a building in Blackheath, South East London.

as life-style demands and increased prosperity create desires in households, for example, to install air conditioning, heated swimming pools and similar facilities that will create an increase in the amount of energy consumed by each household. There will also be predictable increases in consumption of energy caused by population growth, particularly in the undeveloped and developing worlds.

It is worthwhile examining in some detail exactly how the average household uses the energy it buys before we can consider how we can use less energy. It is important, however, to bear in mind that these figures of energy consumption within households do not reflect one critical point. The energy lost in transformation of fuel and delivering is greater than all the energy consumed in the home. So when you consider the amount of energy that a household uses, it is not just the consumption registered on its meters that should be considered but also the huge, unavoidable losses of energy involved in getting the supply into the home that must be taken into account.

First, we shall look at the overall picture. In 1970 domestic households used the equivalent of 37 million tonnes of oil for energy (that is how the government measures these things). Of this 21 million tonnes (57%) was used for space heating, 10 million tonnes (27%)

for water heating, 3 million tonnes (8%) for cooking and another 3 million tonnes (8%) for lighting and other appliances.

By 1997 overall consumption was 44 million tonnes. Of this 57% was still used for space heating, 24% for water heating, 4% for cooking and another 12% for lighting and other appliances.

It appears that when you consider these proportions and take into account energy efficiency savings by insulation generally, households are living more comfortably and luxuriously than they were in 1970. More people keep their houses warmer and use more hot water. They could wear more clothes and wash less, but they do not.

We should approach these figures with some degree of understanding. This is particularly important when considering hot water usage. The figure of 24% of energy in the home being used for water heating includes a great many very small households comprising one or two people. In fact, we believe that an average family with two children of school age probably spends over 30% of its energy expenditure in heating water for household use. A single person household probably spends less than 20%. These figures include bathing, washing, clothes cleaning and general household cleaning water.

Against this background, space and water heating systems have changed developed and improved, providing more luxury for people, with powerful showers, larger baths and hotter homes.

In 1970 only 30% of the housing stock in the UK had central heating. Of those houses that were centrally heated most were heated by coal. By 1998, 89% of the housing stock was centrally heated. Of that 89%, four out of every five houses were heated by natural gas. Over the same period domestic household appliances used 85% more electricity, the greatest increase coming from more people having fridges, freezers, washing machines and tumble dryers and using them more frequently.

For all these improvements, between four and a half and five million households in England (not the United Kingdom as a whole) have to spend more than 10% of their net income (that is more than 10% of what is actually left after rent mortgage payments and tax) on fuel if they want to keep warm. This is a shockingly high figure.

It must be reasonable to assume that the figure required to achieve a decent standard of warmth must be much higher than 10% of income in a good many of these four or five million households, the vast majority of which consist of our elderly poor.

It is difficult to understate the importance of energy in our lives. We cannot live without it. It has assumed a centrality which is fundamental yet commonplace. We have in our government important offices overseeing foreign affairs, economic policy and law and security. Foreign relations, wealth and law and order are all important yet without energy the country becomes vulnerable, poor and weak. When we are governed, it seems that energy is an afterthought, to be considered as part of a senior minister's job, often without sufficient resources. In this way we devalue the tool of energy.

Households in the United Kingdom according to government figures as at 2002 consume 35% of all the energy the United Kingdom creates. Transport consumes 36% and industry only 21%. If you look at the end use of energy the government estimates that 35% is used for transport, 26% for space heating, 8% for hot water, 6% for lighting and appliances and around 25% for process use and other use.

Where does all this energy come from?

Chapter 3

Our traditional energy sources

The starting point in any understanding of energy is to understand the sources of energy and their likely availability in the future.

There are three sources of energy in the United Kingdom. Most energy comes from fossil fuel which is the first source of energy. They are called fossil fuels because they originate from the fossilised remains of animal or vegetable life. These sources are coal (15%) gas (39%) and oil (35%). Fossil fuels are all burnt to create energy, whether it is electricity or heat. Combustion creates the energy, and various gases, the most important of which is carbon dioxide, are created as by-product of the combustion

The second source of energy is nuclear energy and that accounts for 9% of our energy needs. Nuclear energy is sourced from naturally occurring elements which like some fossil fuels are mined but these elements are not burnt. Instead they are combined in a way that generates heat, which is usually used to generate electricity.

The third source of energy is the so-called renewable sources. These virtually all use naturally occurring phenomena, like light, wind and tides, as the energy source to create heat or electricity. This third source in the United Kingdom accounts for only 2% of our energy, with the lion's share of this very small percentage being hydro- electric power which uses falling water to generate electricity.

The balance between these sources has changed over the years. For much of our history wood and peat was a key source of energy and then, as we became an industrialised nation, coal was able to deliver the huge energy requirements of factories. Coal was the only source of gas for many years.

Oil became more and more significant with the rise of the motor car. It will remain a significant source for many years to come. Oil exploration led to the discovery of large quantities of natural gas in the seas surrounding the British Isles. This was found to be suitable for household use and for power station use and over a relatively short period of time coal gas was replaced with natural gas.

The renewable sources of energy are still very much the infant of the energy industry. They do not contribute much to the overall requirement at the moment but for reasons that become clearer in this book, will sooner, rather than later, generate significant amounts of our nation's energy requirements.

The developed world shares the same sources of energy but the proportions vary from country to country, mostly depending upon what is immediately available. The undeveloped world uses traditional sources of energy, mostly in cooking, heating and lighting. These sources tend to be organic fuel and waste, such as waste wood, dung and similar material.

3.1 Coal

Traditionally most of our energy came from coal. Our wealth was founded upon coal. Some 400 million years ago, plant life evolved on earth. 350 to 280 million years ago (a period called the Carboniferous Period after the coal that was laid down during this time) there were masses of vegetation covering the earth. There were many swamps and peat bogs which over the ages accumulated rotting vegetative matter, silt and other sediments. These were very similar to the bogs and swamps of today, but were made up of more primitive vegetation. Violent earth disturbances, called tectonic movements in the earth's crust, buried these swamps and peat bogs, often very deeply. As they were buried, so the vegetation was compressed and heated under enormous pressures.

The pressure and heat and chemical reaction metamorphosed the vegetation into various types of substance which we call coal. Peat was converted into lignite or brown coal, and then, as conditions changed, some lignite deposits were converted to harder, bituminous coals, and under the right conditions converted further into very hard anthracite coal.

This coal, composed of substances that held huge quantities of carbon produced by the vegetation, lay mainly deep underground, the carbon locked temporarily out of the earth's carbon cycle. The sucking of carbon out of the atmosphere by plants (they use the carbon dioxide to produce energy for their growth and existence) prob-

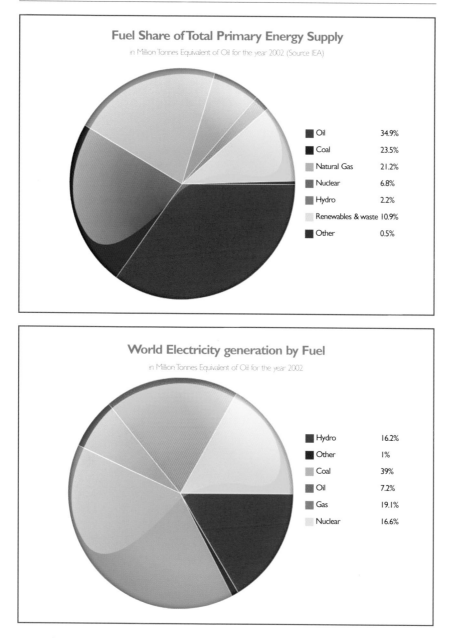

Fuel Share of Total Primary Energy Supply

in Million Tonnes Equivalent of Oil for the year 2002 (Source IEA)

Oil	34.9%
Coal	23.5%
Natural Gas	21.2%
Nuclear	6.8%
Hydro	2.2%
Renewables & waste	10.9%
Other	0.5%

World Electricity generation by Fuel

in Million Tonnes Equivalent of Oil for the year 2002

Hydro	16.2%
Other	1%
Coal	39%
Oil	7.2%
Gas	19.1%
Nuclear	16.6%

Note the predominance of oil, coal and gas – all fossil fuels – in the top diagram. In the lower diagram note that electricity generation is only one part of energy requirement, heat and transport being other important parts. Most of government energy policy is concerned with electricity.

Coal being strip-mined in Montana, USA

ably (and here we can only postulate) cooled the planet sufficiently to enable human life to evolve.

In England in 1700, only 2.7 million tonnes of coal was extracted from simple small scale mines and used for domestic heating. A hundred years later, production had quadrupled to around 10 million tonnes. Then, as man's inventive mind created processes that needed energy for industrial manufacturing, demand for coal grew, so that by 1900, 250 million tonnes of coal were being extracted and used in the United Kingdom each year.

Initially coal was mined by children working long hours in dangerous conditions. Then, in response to social pressure, mining became a man's job. Whole communities came into being in places in Britain where there were rich coal seams. Villages and towns depended solely on the production of coal.

That meant an ever increasing amount of carbon was being "liberated" into the atmosphere. Some of it was discharged as soot and other particulates. Soot is virtually pure carbon. Most of it was discharged as carbon dioxide, a colourless odourless gas that is already significantly abundant in the atmosphere.

1920 was the year when the most coal was mined in the United Kingdom. The coal industry then employed one million two hundred thousand men. Each underground miner mined an average of 200 tonnes a year. By 2001, less than 10,000 miners existed but each miner mined more than 4,000 tonnes a year. Today the UK is producing about half the coal it produced 150 years ago.

In 1970, only 35 years ago and well within the lifetime of most people living in the United Kingdom today, the country produced over 147 million metric tonnes of coal, according to Department of Environment figures. Half of this was burnt in power stations and the heat was used to generate electricity, while significant amounts of coal were still burned in the home for space heating.

Apart from clean air legislation, there was little in the way of environmental protection from the by-products of coal burning, which included not only smoke, carbon dioxide and carbon monoxide, but also other harmful acids that were released into the atmosphere causing acidity in our rivers and lakes.

> Between 1970 and 2002, coal production in the
> United Kingdom fell from 147 million tonnes to
> 29 million tonnes. In 1970, around 137 million tonnes
> of coal was deep mined and only 8 million tonnes
> opencast mined. By 2002 16 million tonnes was deep
> mined and 13 million opencast mined. This present
> production figure is less than the coal actually mined
> in the period of the miners' strike in 1984-85.

The vast majority of current production is now used in power stations, but in 1998 power stations burned about two thirds of the coal they burned in 1970. Our coal production has fallen significantly, but imported coal is now extensively used. In 1970, we imported no coal but by 2002 we imported virtually as much as we produced in our own country.

We have saved money by closing down most of our coal mines and making those in the mining industry redundant, replacing the coal we no longer mine with imported coal mined in other countries. Today, 83% of the coal consumed in the United Kingdom is used for generating electricity. Only 2% is used in domestic consumption. Most of the rest of the coal is used for industrial processes where coke ovens and blast furnaces are needed for iron and steel production, cement and brick manufacturing and glass making.

> Taking the world as a whole, coal supplies 38% of the world's electricity and 23.5% of the global primary energy demand.

Coal can be graded into three types. Steam coal has lower energy content but comprises the bulk of coal used. Coking coal is coal suitable for coke ovens and similar processes. Generally, it has higher energy content than steam coal. Anthracite has very high heat content and this makes it suitable for specific industrial processes where high heat is required and for domestic fuel.

> Most (by far) of the imported coal that we use comes from outside the European Community. 80% comes from just four countries – Australia, Colombia, South Africa and Russia. South Africa sends us 20% of all the coal we consume.

In 1993, we held 45 million tonnes of coal as stock. That would have kept us going for four and a half months of supply. In 2003 we had reduced stockholding of coal to less than three months supply and the trends seems to be that we shall stockpile less and less coal.

Coal production is influenced by international prices. There is also a demand for better quality coal producing fewer pollutants. Most official bodies, including the European Commission, foresee a further decline in the coal market over the coming years. All European coal is now very expensive to produce, compared with imported coal. Although coal production in the European Community in 1998 was about 107 million tonnes, production is reducing year by year at a rate of about 15 million tonnes a year. As European Community production falls, so imports of coal into the Community will increase.

At one time we talked derisively about "taking coal to Newcastle" because the North East had plenty of coal. Now we are taking

Ferrybridge Power Station, West Yorkshire England. Ferrybridge is a coal powered electricity generating station, owned by Scottish & Southern. It has a capacity of about 2GWe. Scottish and Southern are planting trees, hedges and bushes around the power station to encourage wild life and help the environment. The huge cooling towers are used to cool the steam created in the process of electricity generation. This energy is wasted by being emitted into the atmosphere, when it could be used to heat many of the homes in the neighbourhood.

coal to Newcastle from Poland, Columbia, South Africa and even Australia and Indonesia. Because most of the consumption is used by power stations, the utility companies buy, on the whole, high quality coal produced from modern mechanised mines. In some third world countries, children still mine coal, but that is usually on a small scale and child-mined coal is shunned by the utility companies.

The European Commission forecasts that coal imports will actually decrease by about 5% annually – which reflects an overall trend in the United Kingdom and the rest of the European Community to burn less and less coal. We expect this trend to continue and that the environmental costs of coal burning, discussed later, will ultimately prove to be too high for any reversal in this trend.

Coal, mined in China and being loaded by hand at the Yangtze River for distribution.

How much coal is left? There are finite supplies of any fossil fuel and there is always speculation on when these supplies will be exhausted. At the moment the best estimate is that there are relatively large amounts of coal left, about 200 years' worth, if extraction continues at the same rate as present.

It is noteworthy that China mines more coal than the United States and consumes more that it produces. Of the developed countries, only Australia consumes significantly less coal that it produces.

However, lack of supplies or difficulty in extracting coal is not the reason why many commentators believe that in future the importance of coal as an energy source will diminish. There seems to be sufficient coal and modern mining technology is very good at extracting coal economically.

Many countries are seeking to reduce reliance on coal because they regard the environmental burden that the use of coal brings as too onerous. Uncontrolled burning of coal is hazardous in several ways, environmentally. It would have been inconceivable for a sane person to propose that we should control the burning of coal a hundred years ago. Today the World Energy Council has pithily stated

"Coal will need to reduce its environmental footprint" and hardly anyone thinks that this statement is controversial.

Burning coal produces more carbon than burning any other fossil fuel. Many developed countries are looking at ways to burn coal while capturing or sequestrating the carbon by-products and it is expected that as these technologies improve, coal will retain its place as a major source of energy. The less developed world is likely to be less able to do this and we shall explore the tensions that have arisen as a result subsequently.

3.2 Oil

Oil is a very important source of fuel. Like all fossil fuels, it was formed millions of years ago when organisms died and their remains accumulated on the floor of oceans and lakes. Subsequently, these remains were covered by sediment, probably before the remains were fully decomposed in the atmosphere.

The decaying process worked on these organic remains until they were formed into kerogen. Kerogen is a solid, waxy, organic substance that forms when pressure and heat from the earth act on the remains of plants and animals. Kerogen converts to various liquid and gaseous hydrocarbons at a depth of 7 or more kilometres and a temperature between 50° and 100°C. This oil seeped into porous rocks until the oil met a layer of rock that it could not penetrate. As a result, oil bearing sediments formed underground.

Oil is therefore inside porous rocks rather than in underground pools. These porous rocks hold oil rather like a sponge holds liquid. In this form, oil is known as crude oil. It is extracted and refined into many fuel products, including household oil fuel, (kerosene) aviation fuel, and petroleum and diesel oil for vehicles. Oil is also burnt to generate electricity.

When the oil companies started to drill in the North Sea forty or so years ago, they hoped to discover oil. They did find oil, but natural gas discoveries proved to be more important than the oil they found. Nevertheless, they found enough oil to make the United Kingdom an oil producing country in the space of a very few years and although oil was always an important source of energy, the fact that the UK produced sizeable quantities of its own oil, enabled the energy producing industry to be less dependent on volatile regions of the world for its oil.

We first produced oil in the United Kingdom as early as 1919 although our country's companies had led the way in discovering oil in many parts of the world. In 1970, we produced very little oil in the United Kingdom – less than 160,000 metric tonnes – a tiny fraction of our reserves. Within 25 years we were producing 126 million tonnes of oil a year. This amounts to 7.8% of our reserves each year. Oil refineries have always produced more than the country demanded. The actual quantity of product refined has remained relatively unchanged from about 1986 onwards.

For 20 years from 1975, this country consistently estimated each year that its proven, probable and possible oil reserves stand at around 2 billion tonnes. In that same 20-year period we extracted about 2 billion tonnes of oil so that as we extracted, we discovered new supplies. We reported proved recoverable oil reserves of 665 million tonnes to the World Energy Council in 1999. We produced in that year 137.1 million tonnes. That 137.1 million tonnes represented the peak of our oil production.

The figure of 665 million tonnes was the difference between the proven amount of initial recoverable oil that we discovered and the total amount that we had produced out of those discoveries. Clearly the United Kingdom's oil is running out.

There are, it is estimated, another 445 million tonnes of oil where there is a better than 50% chance of extraction, and another possible 545 million tonnes of possible oil reserves where the chances of extraction are still significant but less than 50%. If the price of oil is low, then the cost of extraction can be uneconomic.

Increases in oil prices can make what were unviable oil reserves worth exploiting. The huge oil price increase from around $10 a barrel (a barrel is 35 imperial gallons or 42 US gallons) to (at the time of writing) nearly $70 a barrel is awakening interest of the oil extraction companies in fields that they had previously written off as being uneconomic to exploit.

There is much debate about the true extent of the reserves of oil left in the world today and it is particularly hard to judge "proven" oil reserves given the fact that the likelihood of extraction depends upon the price of oil which has fluctuated greatly in recent years. Oil has particular economic importance and must be a finite resource. Some people think that all of the oil-bearing parts of the earth have been discovered and that the governments of countries that have oil reserves have tended to over estimate the amount of

An oil rig in the North Sea. Most of the North Sea oil has already been ˏ converted into energy.

the reserves available. They believe that future discoveries will be small. This leads analysts to conclude that oil production is at its optimum and will steadily decrease.

Other analysts think that it is important to have regard to the technical advances in discovering and extracting oil. They also point out that the fact is that small amounts of oil that are initially discovered in a particular field often evolve into much larger amounts. In 1979, the Cognac Oil rig drilled oil 312 metres below the sea. By 1999, oil was being extracted 1,853 metres below the sea by the Rocander rig that floats on the surface. These kinds of improvements enable more oil to be found and extracted.

The issue of oil reserves and how much "guess work" exists in this area was brought sharply into focus by Shell. Shell is the third

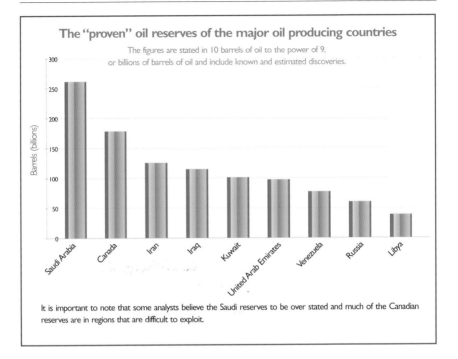

The "proven" oil reserves of the major oil producing countries

The figures are stated in 10 barrels of oil to the power of 9,
or billions of barrels of oil and include known and estimated discoveries.

It is important to note that some analysts believe the Saudi reserves to be over stated and much of the Canadian reserves are in regions that are difficult to exploit.

The "proven" oil reserves of the major oil producing countries. It is important to understand that some analysts believe that the Saudi Arabian reserves are over stated and that much of the Canadian reserves are in regions that are difficult to exploit. The figures are in billions of barrels (10 to the power of 9).

largest oil company in the world. In January 2004, it announced that part of its assets, its proven oil reserves, had been overstated by 20% and that it was revising its proven reserves downwards by this amount in a one-off exercise. However, by May 2004, there had been a further series of announcements by Shell all of which had the effect of reducing the company's proven oil reserves.

Although this news greatly affected the oil market and the stock market, it actually probably has greater significance in trying to understand the world's energy capability.

There are differences in the ways that the US stock market requires oil reserves to be counted from the rules of European stock exchanges concerning the proven oil reserves. This confusion of rules in the financial markets is added to by the ways in which each oil-producing country

The state of the art Leuna Refinery, in Germany. Mineral oil products have been manufactured in the new refinery since 1997. The refinery in Leuna is able to turn about 10 million tonnes of crude oil, mainly coming from Russia, into fuels, heating oil, liquid gas, bitumen, naphtha and methanol.

measures its reserves. This leads us to question how accurate the "proven" oil reserves of the world are.

We have seen that reliance on the world's corporations cannot be 100%; we shall examine the extent to which we may be able to rely on a country's estimates of "proven" reserves by considering the reliability in the proven reserves of the OPEC countries. These are: Algeria, Indonesia, Iran, Iraq, Kuwait, Libya, Nigeria, Qatar, Saudi Arabia, United Arab Emirates and Venezuela. This cartel of oil producing nations accounts for around 40% of the world's oil production and while not controlling the market, it does influence it greatly because its member countries do control around 55% of the oil that is traded internationally.

In 1982, OPEC members agreed to defend what was then a lowering of the oil price by restricting supplies. They agreed to restrict

Oil burning as a result of war, in Kuwait.

supplies by reference to a production quota system. In 1988, OPEC changed their rules; they no longer rationed oil on the basis of their production, but did so by reference to their proven oil reserves. At the same time, most of the OPEC countries (and all of the very large producers) substantially increased their estimates of reserves; overnight the amount of proven reserves of 40% of the world's oil had risen by over 50%.

It may well be that some proven reserves were previously under-stated, but analysis of OPEC's figures by Professor Deffeyes of Princeton University indicates that OPEC members have less oil than their figures claim, possibly by as much as 17%.

If the existing figures for world oil production are accepted, then the reserves will be exhausted in 37 years if the rate of oil consumption remains the same. If the reserves are adjusted to take account of Professor Deffeyes thinking, then we have only 27 years of oil left. If the amount of reserves is adjusted upwards to take account of the oil producing companies' (post Shell) estimates, then we have 41 years of oil left.

This uncertainty does not help the understanding of energy issues. However, what is probably incontrovertible is that the consumption of oil will increase whether production increases or declines.

The world is producing around 28 billion barrels of oil a day but only discovering 7.42 billion barrels a day. However those discoveries are measured, they fall far short of production and consumption. 2003 was the first year in living memory when no significant new oil was discovered. It does not seem conceivable that as the Asian economies of China and India grow, demand for oil will remain static. It seems almost certain that while the oil may not run out in our lifetimes, it will become increasingly expensive and in shorter supply. We shall have to learn to use a great deal less of it.

3.3 Gas

Natural gas is an important fuel in the home, in industry and for generating electricity. In 1970, the United Kingdom produced 11.1 billion cubic metres of natural gas – less than half of 1% of our probable and provable reserves of gas at that time. Within twenty-five years, we were producing nearly 70 billion cubic metres a year – which was just about 4.45% of our reserves.

Most of the gas we use these days come from the down formations laid down millions of years ago. Natural gas is a fossil fuel, created by the same processes that create crude oil. As with oil, organic remains are trapped in sediments and compressed at high pressure for long periods at higher temperatures. Lower temperatures (relatively) form more oil than gas. Gas is usually found in or near oil bearing rocks at depths of between one and two miles below the surface.

The process of gas formation underground at high pressure creates what is known as thermogenic methane. The pressure and temperature underground is believed to break down the carbon in the organic matter. The gas created is almost all methane gas, a colourless odourless substance made up of carbon and hydrogen. There is often a small proportion – not more than 20% - of ethane propane or butane and a small amount of carbon dioxide, although the precise composition varies and that it why the calorific value of gas (the amount of energy that can be derived from it) varies not only from place to place but also from time to time when natural gas is piped into a national network.

Natural gas can also be formed when micro-organisms act upon organic matter. This process created biogenic methane. These processes occur when the micro-organisms break down the organic matter on or near the surface of the earth. The methane produced is mostly lost to the atmosphere, although from time to time it can be trapped in underground caves and exploited for energy use. Landfills usually produce large amounts of methane as waste matter decomposes. Important, relatively new technologies are seeking to harvest landfill gases and use them commercially.

Methane is also formed through a chemical reaction caused when hydrogen and carbon, lying deep within the earth, rise through the rocks and react with minerals. This process creates biogenic methane.

When gas is formed by a thermogenic or a biogenic process, it is usually held underground under pressure. When a drill penetrates the underground gas, the pressure is released and gas rises to the surface.

In ancient Greece, natural gas seeped from the hillside of Parnassus at a place now called Delphi and burned constantly. The Greeks built a temple over the place where the flame burnt. The burning flame was used by the priestesses of the oracle to inspire their predictions of the future.

In industrialised Britain, gas was formed from coal and used for lighting from about 1875. There was initially no long pipe infrastructure so gas tended to be used where it was found or made. In 1885, Robert Bunsen invented his burner which, by mixing gas with air, created a flame and heat consistency that was well adapted to cooking use. Eventually, most industrial countries built a gas network of pipes so that gas could be transported around the country and used for heating, cooking and water heating. Gas lighting became virtually redundant when electric lighting was invented.

Today, natural gas is the source of 39% of our energy. Our own reserves are dwindling and in the first decade of this century a major source of our gas will come from Norwegian fields. That will not be sufficient for the United Kingdom. We will need to buy gas from many other parts of the world. The most likely suppliers will be Russia and North Africa as well as politically unstable countries in Latin America, Central Asia and the Middle East. By 2006, the United Kingdom will be a net importer of energy raw material including

This rig, in the special protection area of the Humber estuary, near Killingholme delivers both oil and natural gas.

natural gas. 30% of natural gas is burned to generate electricity in our power stations but households directly consume 35%. They burn gas in their boilers for space heating and hot water.

> The definitions of our gas (and oil) reserves are important. The UK government defines proven reserves as known reserves which have a better than 90% chance of being produced. Probable reserves are known reserves which are not yet proven but which are estimated to have

a greater than 50% chance of being technically and
economically producible. The government's estimate of
proven, probable and possible gas reserves is actually
increasing year on year notwithstanding the fact that we
have extracted over one trillion cubic metres of gas over
the past 20 years.

Natural gas has wholly replaced coal gas, that is to say, gas produced from coal. There is a sophisticated network of supply pipes throughout most of the country and the vast majority of households have access to piped natural gas. Northern Ireland at present has almost no natural gas although there are plans to connect it to the mainland gas network.

For those households, particularly in rural areas, which do not have access to piped natural gas, there is liquid petroleum gas and calor gas that can be delivered to storage tanks at the home. Sophisticated remote monitoring devices enable gas tanks to be filled without the consumer having to make a telephone call; the supply is now virtually as seamless as a piped supply.

Natural gas burns much more cleanly than oil or coal. This is because it has less carbon and more hydrogen. When it burns, it emits less carbon dioxide per unit of heat provided. If gas is not burned but released as methane into the atmosphere, its effects on global warming, as we will see, are very deleterious but burning it is a cleaner option than burning any other fossil fuel.

Natural gas has fewer particulates too; it will release 5% of the particulates that oil releases and less than 1% of the particulates that coal provides. It releases almost no sulphur dioxide (which creates acid rain) and very few nitrous oxides. In environmental terms, it is far cleaner than oil or coal. Inevitably, it does release carbon dioxide like all burning.

The world's largest reserves of natural gas are in Russia,
which estimates it has 47 trillion cubic metres. Iran has
over 24 trillion cubic metres and the other OPEC
countries also have significant reserves, especially in the
Middle East. The Americas and Africa together appear to
hold less than 17% of the world's supply and populous
Asia only 11%.

World gas reserves were estimated by Oil & Gas Journal at 6,076 trillion cubic feet as at January 2004. There has been a year on year increase in the estimates of gas reserves for the each of the preceding nine years. Consumption of natural gas is also increasing year on year and present estimates assume that we have more than sixty years' supplies of natural gas. This level of reserves compares very favourably to oil reserves. The United States Geological Survey postulates that there are over 4250 trillion cubic feet of undiscovered natural gas.

The most significant fact about gas reserves lies in the regions where they are located. Over 75% of the reserves are in the Middle East and Russia, and if, as some commentators believe, oil becomes harder to find and more expensive to produce, those countries rich in natural gas will exploit the reserves that they control for their own benefit.

3.4 Hydro Power

Ultimately the energy in falling water derives from the sun – hydro-electric energy is another form of solar power. Energy contained in sunlight evaporates water from the oceans and deposits it on land in the form of rain. Differences in elevation result in rainfall run-off which allows some of the solar energy to be converted into hydro-electric power.

Falling water has been used as a power source for thousands of years: water powered clocks in ancient times and then mills to grind corn and for other applications. In 1882, moving water was used to power a waterwheel on the Fox River, Wisconsin, and electricity was generated. Shortly after the Fox River experiment proved the viability of hydroelectric power, the first of many generating plants was built at Niagara Falls.

Early hydroelectric plants were very reliable and more efficient than the oil and coal fuelled plants of the day. As the price of oil and coal fell, and electricity demand soared, small hydro plants fell out of favour. Most new plants were huge and involved major dams.

These dams flooded vast areas of land. This in itself causes humanitarian and environmental concerns. Until recently, most people believed hydroelectric power to be safe, clean and environmentally positive. There is no emission of carbon dioxide or sulphur dioxide and there is no risk of radioactive contamination.

However, some studies reveal that large reservoirs created by flooding do produce harmful greenhouse gases. The decaying vegetation under the water, produces, it is believed, the same sort of quantities of greenhouse gases as those produced from other sources of electricity.

Hydroelectric power is generated by water flowing through turbines. Great falls are needed to make the generation effective and often vast volumes of water are needed to make the production of electricity commercially viable. When hydroelectric power is actually generated there is no atmospheric pollution, no radioactive waste and certainly this form of power can be pollution-free, or so it may be thought.

One problem lies with the large volumes of water that must be used. Many places are too flat to make hydroelectric generation effective. In some countries the rain falls in the wrong places and so large areas of water have to be collected and dams built for this power to be generated, and therein lies the difficulty.

Flooding large areas of land has environmentally damaging effects. First, there is the obvious loss of wildlife, vegetation and people's homes. Second, and less obviously, some scientists calculate that the decomposition of vegetation caused by flooded lands in dam projects gives off substantial amounts of carbon dioxide which, of course, contribute greatly to acid rain and to global warming.

Canada is a country that acquires more of its power from hydroelectricity than any other source. Canada has plans to flood large areas of land in Quebec to create dams for hydroelectric generating plants. It has already flooded over 10,000 square kilometres of land in the La Grande project in the James Bay region of Quebec and if further plans are carried out, then the eventual area of man-made lakes caused by flooding in Northern Quebec will be an area that is larger than Switzerland.

It is feared that by flooding such a vast tract of land not only will greenhouse gases be created (which will be equivalent to those produced by burning fossil fuel) but there will also be a loss of rare ecosystems, flora and fauna. In other places, large dams have in addition destroyed the homes, communities and ways of life of peoples who have been indigenous in these areas for thousands of years. These are the hidden ill-effects of hydroelectric power.

Of course, hydroelectric power, which does not require flooding, where the turbines can be driven by the run of water without dams

Claerwen dam, in the Elan Valley, Wales. This dam was built to supply water. As a result a school, a church and many homes and small farms were lost; bodies from the local graveyard were exhumed.

or reservoirs, would not be a source of greenhouse gases nor the other ill-effects described.

I should also mention the short-term problems of hydroelectric power. Damming a river can alter the amount and quality of the water in the river down stream and can prevent fish from travelling upstream to spawn. This can be (partially) resolved by fish ladders – but there is a limit to how high fish can climb. In addition, silt, normally carried downstream naturally by the flow of the river to where it fertilises the estuary lands, is no longer carried down to the river mouth where a dam intervenes. It slowly accumulates in the reservoir until it eventually decreases the amount of water that can be stored. For example, after only four years of being in full operation, the Sanmen Gorge dam, on the Yellow River in China, had lost over 40% of its water storage capacity and 75% of its 1,000 MW power capacity, due to the sediment build up in the reservoir.

Rotting vegetation under water becomes a rich habitat for bac-

The Aswan Dam, in Egypt

teria. Bacteria can change the character of mercury, present in rocks under the reservoir, making it water-soluble. Mercury then accumulates in the bodies of fish and, of course, it is dangerous to eat fish with mercury in their bodies. In the La Grande dam project, mercury has been found in both the fish and the water where none was found before the project started.

The bacteria also change the quality of the reservoir water. New forms of bacteria can develop in these man-made lakes that can pose serious health hazards.

Finally, all large bodies of water, whether man-made or not, do influence the climate of the locality surrounding them. Water from reservoirs evaporates and as a result, humidity levels tend to be higher, causing more fog than normal. In tropical areas man-made lakes are thought to disrupt the convection cycle and ultimately reduce cloud cover.

There are some famous hydroelectric projects which, when they were conceived, were thought to have no significant polluting consequences but, as always, time presents the bill that must be paid.

Egypt built a 360 foot dam across the River Nile near Aswan. It is nearly two and a half miles long and stores one and a half billion cubic feet of water. It produces electricity and has helped control the flooding of the Nile, but has uprooted large numbers of the indigenous population and submerged ancient monuments. On the River Paraná between Paraguay and Brazil the Ituaipu Binacional power station feeds off an even bigger dam – nearly twice the size of Aswan.

Turkey, a growing and thriving country, needs electricity and energy and has turned to hydroelectric power. Seven multinationals and eight governments, including that of the United Kingdom, were planning to build the Ilisu dam in south-east Turkey. The dam would displace 78,000 people from their homes and many farmlands (and accordingly livelihoods) will be lost. A dam will flood the medieval town of Hasankeyf where artefacts pre-dating biblical times are thought to exist. Clearly the environmental damage will be enormous. Critics point out that the affected persons are all Kurds and that, when considered with other existing dams, the flow of the ancient River Tigris will be controlled by Turkey, to the detriment of Iraq and Syria.

This project was initially intended to be undertaken by British civil engineering companies with underwriting support from the UK government. It is thought that local pressure campaigns against the dam led to the withdrawal of British interest but it now seems that Siemens is intending to revive the project.

One huge dam project in the course of construction is in China on the River Yangtze. China has conceded that the project is likely to cause serious environmental damage. The project will raise the water level by up to 250 feet over a 200 mile stretch of the river and will create a reservoir of 252 square miles. So far more than 100,000 people have been moved from small farms and towns close to the river. By 2003, another half a million people will have to be moved and another half a million three years later. China plans to rebuild villages and towns wherever possible further up the hillside. Critics argue that the higher slopes are mostly barren and the area is too isolated for businesses to be established there.

Like all major projects there are arguments for and against. The dam will help prevent flooding downstream and this is of extreme importance to the millions who live there. It will generate abundant electric power for central China, supplying more than ten large cities with electricity. However, most of the flooding downstream appears

The Three Gorges dam being built in Sandouping, China.

to be caused by tributaries rather than the Yangtze itself, and there will be, at the end of the project, more than one million people who have been forced to leave their homes.

> Fifty or sixty years ago the United States considered that dams were mainly beneficial. After all, the Hoover Dam made the City of Las Vegas possible. Today there is a lobby for returning rivers to their natural ways. Policymakers are seriously considering pulling down hundreds of dams in states such as Wisconsin and Pennsylvania. Between 1990 and 1998 more than 200 dams were dismantled.

Dams in many cases no longer serve much purpose and do a lot of harm by altering the natural flow of rivers and affecting the fish and other species that depend on them. There are about 75,000

The Hoover Dam

dams in the United States, but only a few thousand are likely candidates for removal in the near future. Demolishing even a dozen medium-sized dams could consume billions of dollars and decades of effort, and come at a huge cost to the environment because of the enormous amounts of silt and sediment trapped behind the structures. Engineers and ecologists agree that dismantling dams, particularly the big ones that impound water in reservoirs, is a delicate and difficult operation. The silt and stones trapped behind the structures can clog the ecosystem downstream. In many cases, if the sediment were allowed to flush past the dismantled dams, it might smother the rivers, wiping out the very habitat the dam removals were undertaken to protect.

Some hydro power plants do not use dams. Instead, a section of a river is diverted through a channel or penstock and turbines are placed in the channel to use some of the river's energy to generate

The Glen Canyon dam in Arizona, showing the hydro electric plant.

electricity. This dam-free method was used in Alaska at the Tazim-ina project. The environmental problems associated with dams do not arise when this method is used but it is not always commercially viable to divert a section of a river.

Using water to store energy is also becoming more widespread. When electricity demand is low, the surplus current can be used to pump water from a low reservoir to a higher one. When electricity demand is high, the water in the higher reservoir can be fed through turbines generating electricity as the water falls back into the lower reservoir. This is one of the few ways that electricity can be "stored".

Hydro power is a very important way of making electricity for the world. It provides around 19% of the world's electricity amount-ing to over 2,650 TWh a year. It is not such an important energy source for the United Kingdom. The natural gentle contours of most of the United Kingdom do not lend themselves to producing more hydroelectricity than we produce at present. In addition, people want to preserve the environmental character of our small islands and do not take kindly to the prospect of building large dams across valleys of outstanding beauty.

The actual figure for renewable sources of electricity generation in the United Kingdom in 2002 was only 3% of the total amount of generating capacity – a modest 1.5 gigawatts. There is probably some scope for improving this by using small scale local hydro schemes.

3.5 Nuclear Power

Another major source of the power we consume is nuclear energy. There are over a dozen nuclear power generating plants in the United Kingdom and the generation of electricity by nuclear power started as early as the mid 1950s. The United Kingdom Atomic Energy Authority's figures show that the nuclear industry has grown into a major electricity generating force.

> By 1997 nuclear power generated 26% of our country's electricity production, compared with 33% for coal and 29% for gas. Today it generates around 20% of our electricity needs and is the source of 9% of our primary energy demand.

Nuclear power plants are being closed in the United Kingdom but the government does not rule out building new nuclear power stations if we fail to meet our carbon emission targets.

Nuclear energy is a very new source of power. It was only a theoretical possibility before the Second World War. In early 1939 uranium fission – the splitting of the uranium atom – was accomplished and considered practicable. One scientist, Hungarian born Leo Szilard, was concerned that the United States was oblivious to the possibilities that this discovery opened. He was without any influence but he had a highly respected and influential friend, Albert Einstein.

Szilard persuaded Einstein to write to President Roosevelt warning of the dangers. On 2nd August 1939 Albert Einstein wrote to President Roosevelt warning him that nuclear energy was more than a theoretical possibility. His letter was jointly composed with his friend Szilard, but Einstein's signature gave its warning authority, made starker when shortly after the letter was delivered the Second World War broke out.

> *"In the course of the last four months it has been made probable – through the work of Joliot in France*

> *as well as Fermi and Szilard in America – that it may*
> *become possible to set up a nuclear chain reaction in a*
> *large mass of uranium, by which vast amounts of*
> *power and large quantities of new radium-like*
> *elements would be generated. Now it appears almost*
> *certain that this could be achieved in the immediate*
> *future. This new phenomenon would also lead to the*
> *construction of bombs, and it is conceivable – though*
> *much less certain – that extremely powerful bombs of*
> *a new type may thus be constructed."*

Einstein's letter led first to the United States and then the Allies building the nuclear bombs that were ultimately exploded on Nagasaki and Hiroshima in 1945. The German government also researched this possibility but had their efforts hampered by allied bombing of critical installations. The allied research that was carried out also led to nuclear energy being developed as an energy source, just like coal and oil, in its own right. The process that Einstein warned of was the fission or splitting of uranium atoms, generating energy in the form of an explosion.

Nuclear energy in fact releases the energy stored in uranium and converts it to heat and then electricity, not into an explosion. Uranium is placed in a core, known as a reactor because it is part of the operation that causes the uranium to react. Uranium has 92 electrons that orbit the nucleus of each atom. As the uranium is drawn out of the core, graphite is used to slow down the movement of the uranium neutrons. Slowing the neutrons down enables two or three of the 92 neutrons to be released from their orbit. One of these displaced neutrons hits the nucleus of another uranium atom. That causes the atom to split. This sets up a chain reaction of more neutrons being displaced hitting more and more atoms so that lots of atoms split.

As they split, the atoms are effectively exploding in a small and controlled way. That releases tremendous amounts of heat – known as kinetic energy – which is used to turn water into super hot steam which drives a turbine, generating electricity. The used steam is cooled, usually in the sea (the water going into the sea is usually around 10 degrees Celsius warmer than it was when it started). The electricity from nuclear plants usually constitutes the base load of the United Kingdom's electricity supply, because nuclear power stations take longer to "turn off" than other power stations.

Dungeness nuclear power station at night. The advanced gas cooled reactor type "B" is on the left hand side and the type "A" Magnox is on the right.

There are, of course, other installations: British Nuclear Fuels owns and operates eight Magnox nuclear power stations in West Cumbria, Southern Scotland, Essex, Kent, Somerset, Avon, Suffolk and in Anglesey.

Calder Hall in West Cumbria was the world's first commercial scale nuclear power station and was officially opened in 1956. It now generates enough electricity to supply a city the size of Leeds. Between them, the Magnox power stations alone provide around 8% of the UK's electricity.

Properly designed and maintained nuclear generating stations produce electricity without producing emissions. Indeed, the lack of carbon dioxide, carbon monoxide, smoke and other emissions is an important factor in favour of nuclear power stations. Nuclear power does not create by-products of smoke, carbons and noxious gases because no burning process is involved and therefore no carbon is created. It does cause large amounts of heat to be deposited in water, usually the seas which are warmed by the plants.

Nuclear energy is probably the cleanest form of non renewable power in the conversion of matter into energy but it is not without considerable disadvantages. Were it not for the problems of safety and radioactive waste, nuclear energy would be an almost ideal source of power. Relatively small quantities of uranium are needed to generate huge amounts of electricity.

Radioactivity is a random process where the nuclei of atoms disintegrate spontaneously and as they disintegrate they emit rays. This process is naturally occurring (in sunlight, from rocks and from trace substances around us). In 1903 Becquerel was awarded the 1903 Nobel Prize for physics when he showed that uranium emitted rays without any external energy source. Although the phenomenon was discovered by Henri Becquerel, the term radioactivity was invented by Marie Curie who saw that after refining uranium ore the residual material was more active than pure uranium. She and her husband Pierre were also awarded the 1903 Nobel Prize in physics.

The most important work, however, was done by Ernest Rutherford at the University of Manchester in the early part of the last century. Rutherford invented the language to describe the theoretical concepts of the atom and the phenomenon of radioactivity. Rutherford startled the world by proving that uranium became a different element through the process of radioactive decay. He had shown that transmutation of elements actually occurred at a time when virtually all thinking had dismissed the turning of one element into another to the realm of alchemy, not science.

He too won the Nobel prize in 1908; the following year he experimented with bombarding thin foils of gold with particles. He noted that some of them bounced back " "as if you fired a 15-inch naval shell at a piece of tissue paper and the shell came right back and hit you."

From this, Rutherford deduced that the atom's mass must be concentrated in a small positively-charged nucleus circled by electrons inhabiting the farthest reaches of the atom, rather like planets whirling round the solar system. Rutherford's genius and the ground breaking work and thinking he did at Manchester enabled all nuclear technology to follow.

Nuclear processes, in power plants like the Magnox plants, generate a significant part of our electricity. Nuclear energy generation does not produce any carbon or sulphur gases as a by-product; it generates no black smoke. However, nuclear energy is not free from carbon production when it is considered over its whole process. Carbon dioxide is released in very process up to (but not including) actual fission of the atoms in the reactor. There is the mining, milling and enrichment of the fuel, all of which produce carbon dioxide, and the construction of the power stations and the handling of the spent fuel.

Low grade ores need a lot of energy to mine and that releases much carbon dioxide. Only Australia and Canada have high grade ores. At ore grades of below .01% for "soft" ores and below .02% for hard ores more carbon dioxide is released than in an equivalent gas fired power station, according to Jan Willem Storm van Leeuwen, a leading consultant in energy systems. He suggests that ores of a grade of those found in India, if used, are likely to create a net energy loss, rather than a gain, taking all matters into account.

Subject to the above caveat, nuclear energy should not contribute to global warming. It may be marvellously clean except for one factor. It does produce as a by-product: radioactive waste.

As clean and carbon free as the process may be, the dirty by-product needs to be stored for thousands of years and that has carbon consequences. The best known storage involves the building of large concrete underground storage facilities. When cement is produced carbon dioxide is also produced as a by-product and the global cement industry knows of no way to avoid it. Around 1.4 billion tonnes of CO_2 are produced by the cement industry each year, which is a startling 6% of the man made sources of carbon dioxide in the atmosphere.

When you consider the actual cement used not only in nuclear power station construction but also in vast nuclear waste storage facilities it is simply not right to consider nuclear as a carbon free source of energy.

Some sources have concerns that the nuclear industry is unsafe. We shall discuss this later. There may be accidents or accidental leakages of radioactive material. There is a debate as to whether there are significantly higher levels of leukaemia around nuclear reactors. Leaving aside these real concerns, there is one problem that must be addressed in every nuclear operation – that of what should be done with the radioactive by-product of used fuel.

Nuclear installations at Dounreay, Windscale (Sellafield), Harwell, Culham, and Winfrith are all being dismantled under the auspices of the Atomic Energy Authority. At Winfrith, in Dorset, the site opened in the mid-1950s and its main reactors were closed in 1991. Decommissioning is still going on and it is projected that the site will be safe enough to be used as a business park by 2025. That projection is likely to be overoptimistic.

The real problem is that there are only two basic options for used fuel – either direct disposal or reprocessing.

Before reprocessing, the potent used fuel must be stored for some five to ten years at reactor sites or reprocessing plants. During reprocessing, uranium and plutonium are separated and various intermediate- and high-level wastes are generated. Here we also find very long-lived, high-level, liquid waste. This is mainly the fission products and actinides other than uranium and plutonium.

Most potent of all used fuel is long-lived high-level solid waste. The whole fuel assembly of the reactor is treated as waste. It is stored and cooled until its radioactivity and heat output have reduced enough to simplify handling, before it is made ready for packaging and final disposal. These storage and subsequent conditioning operations (called reprocessing) generate intermediate- and low-level waste.

There is also long-lived intermediate-level liquid waste, which arises from the various separation processes and which must be safely disposed of, together with intermediate-level solid waste, such as fuel claddings. In addition, there will be a variety of low-level wastes to dispose of.

Liquid high-level waste must be stored in cooled tanks for a number of years before being solidified by a vitrification process. Steel containers holding the vitrified waste are then stored, typically for

Nuclear waste embedded in glass and cement for storage.

30 to 50 years, in air-cooled vaults. This is not the end of the process of making the high-level waste safe, but, sixty years after the waste was created, the process is merely beginning. The waste will be "geologically disposed of" by which the nuclear industry means that they will bury it in concrete structures in what they hope are stable rock formations for thousands of years.

The nuclear industry will have evaluated the safety of the disposal site and the storage system, and they will assure us that their evaluation is 100% accurate. Without being over cynical, it seems to us that this long process must involve certain significant risks, no matter how much care is taken. Thousands of years amount to a long period of time – far longer than recorded history.

Low- and intermediate-level liquid wastes are also created. These effluents (but not high-level waste) are sent to a waste treatment plant where they are stored in different tanks, depending on their nature. Two types of treatment can be used for decontamination: chemical precipitation or evaporation. The radioactive sludges, flocs, and evaporator concentrates that the nuclear processes create are solidified by being cast in cement, resins or bitumen.

Why is this nuclear by-product so dangerous? The rays emitted are invisible, odourless and silent. They have an ability to ionise living cells and this ionisation process brings about change to the living cells. The cells may die or its DNA may be altered. Cell death at a massive level ends life and DNA changes frequently bring cancers to life.

Many people are rightly uneasy about the storage and disposal of these dangerous by-products. Others feel strongly that the safety issues are not sufficiently addressed in the nuclear industry. The volume of highly radioactive irradiated or spent nuclear fuel has massively increased. Despite the huge scientific and economic input of a civil nuclear programme spanning forty years, there is still no accepted technique for disposing of the waste that inevitably results from the operation of nuclear reactors. There is now more than 140,000 tons of spent fuel from nuclear plants being stored in pools of water at reactors. The amounts stored are increasing daily.

When things go wrong it is catastrophic. It is not too long ago that the accident at Chernobyl not only killed many people in the then Soviet Union but the release of radioactive material was measurable in livestock all over Europe.

It is worth considering the damage caused by the events at Chernobyl in some detail because it will assist us in evaluating the real dangers of the nuclear industry.

On 26th April 1986, a nuclear reactor at the Chernobyl Nuclear Power Station, located 60 miles north of Kiev, in what was then the Soviet Union but is now the Ukraine, accidentally exploded. The accident occurred during an experiment designed to improve safety.

In all nuclear reactors a great deal of heat is created. Overheating is dangerous and so the power station engineers are constantly trying to find out whether they can improve safety and reduce the risk of overheating the nuclear core.

Chernobyl was designed to have its core cooled by water and its emergency core cooling system, designed and installed to prevent overheating, was powered by electricity. The power station engineers were rightly worried that if the electrical supplies to cooling pumps were to fail, the core might overheat. They had been conducting a series of experiments to discover the best way to prevent overheating.

On 26th April in the course of one of these experiments the engineers operated the nuclear reactor outside its normal range. They switched off the emergency core cooling system (for eleven hours) and operated the reactor at less than full power in order to establish whether the main pumps, powered by the residual electricity

A cow grazing in Georgijevka. The sign warns of radioactive hazard.

from a generator that was being slowly run down, would be able to cool the core.

When these types of power stations work at low output, more steam than water gathers in the fuel tubes and consequently the number of neutrons available for the chain reaction actually increases. This flaw was overlooked at the design stage. Although the engineers were reducing reactor power, the low flow of cooling water boiled, producing more steam, thus increasing the quantity of neutrons and causing the power to rise.

Eventually, as things started to go wrong, they shut down the generators. The cooling pumps powered by the generators slowed. This reduced water flow in the core, producing more steam. This caused the number of neutrons in the chain reaction to increase further and reactor power actually increased.

Two non-nuclear, but chemical, explosions occurred as a result of the interactions between steam and the overheated fuel elements. The force of the explosions was equivalent to 200 tonnes of explosive. The reactor 'lid', (it weighed more than 2,000 tonnes) lifted off and tonnes of nuclear material escaped into the atmosphere.

The control room in fast breeder reactor at the Beloyarskaya nuclear power plant.

Nuclear Energy Agency published the results of further research that indicated the release was actually three times greater than previously estimated.

> As at 31 December 1995, there were some 430 commercial nuclear power reactors operating in the world with a collective nominal capacity of nearly 340 gigawatt (GW), producing about 17% of global electricity.

Seven years before Chernobyl there was an accident in another nuclear power plant. This time it was not poorly designed or inadequately regulated. It was operated in the most technologically sophisticated and power-greedy country in the world – the United States. The accident happened at a pressurised water reactor at Three Mile Island nuclear power station near Harrisburg, Pennsylvania. The reactor was a new one and it overheated (or in the words of the nuclear industry "suffered a loss of cooling at the reactor core"), which resulted in partial melting of the core.

To avoid failure of the containment building, the fission product gases were released into the atmosphere 48 hours after the accident and the hydrogen over a period of a few days. The area around the plant was temporarily evacuated while this was done. No general evacuation was carried out but it remained a possibility for several days. The accident caused no injuries and many highly regarded epidemiological studies conducted between 1981 and 1991 have found no measurable health effects to the population in the vicinity of the station.

Today, the Three Mile Island reactor is permanently shut down, with the reactor coolant system decontaminated, the radioactive liquids treated, most components removed and the remainder of the site being monitored. The owner, General Public Utilities Nuclear Corporation, says it will keep the facility in long-term storage until the operating license for the Three Mile Island power station expires in 2014, at which time the whole plant will be decommissioned.

Closer to home there have been scares and problems with nuclear safety at power stations in the United Kingdom. In October 1957 a fire occurred at Windscale in a military reactor used to produce plutonium for Britain's nuclear weapons programme. Overheating caused it and, as a result, releases of fission and activation products into the atmosphere occurred. The fire was eventually put out on Friday 11th October 1957 by simply flooding the reactor with water. There were tiny, but measurable, increased cancer risks – particularly thyroid and lung cancers. Compared with Chernobyl the release of dangerous material was small. Later the authorities renamed Windscale "Sellafield".

The additional risks from the Windscale fire are small in comparison with normal cancer rates. Taking into account all measurements, the total fatal cancer risk from the Windscale fire to the most exposed individual is only marginally higher than the normal fatal cancer risk for a member of the UK population, according to the nuclear industry.

For many years now there has been a steady decline in the fortunes of the nuclear industry. We have seen the cancellation of nuclear power programmes and reactors around the world. Today, in Western Europe only France has any reactors under construction, while in Central and Eastern Europe only a handful of reactors are being built. In Asia reactor programmes are being slimmed down and cancelled. After a review of the privatisation of the British nuclear power industry, it was concluded that there was no economic justification for public funding to build any new reactors. In December 1995, Britain announced the cancellation of its two proposed nuclear power stations. In the next decade this downward trend is likely to continue, and as the true economic and environmental costs of decommissioning and radioactive waste management are discovered it could even accelerate more rapidly.

In 2005 Nuclear Power Stations were being suggested as a way of reducing carbon emissions. Many famous people living in areas of outstanding natural beauty preferred the nuclear option to wind farms for environmental reasons. It is difficult to know how this debate will resolve itself because nothing has been ruled out.

In April 2005 another incident happened at Sellafield; it was classed as a level 3 accident on the International Nuclear Event Scale. Chernobyl was level 7 and Three Mile Island level 5 so while it was nowhere near as serious as these accidents it was nevertheless worrying. No one was injured or damaged by the leak, which was fortunately discovered in time.

A small thin pipe fractured, probably as a result of metal fatigue, leaking nitric acid on to the floor of a concrete-lined sealed cell which was inaccessible to staff. It is thought that the failure may have been occurring for nine months before it was discovered. Twenty tonnes of uranium and 160 kg of plutonium were discharged into the cell. No one noticed.

Some take the view that the nuclear industry has had almost 50 years to prove that nuclear technology is safe, clean and cheap and has failed to do so. It boasted originally that it would produce electricity "too cheap to meter" but now it is clear that the environmental and economic costs of nuclear power may well be too expensive to afford.

When the industrial revolution took place most people could see the benefits that the new machinery invented would bring to the lives of people. Luddites took to smashing machinery because they objected to the changes which for them meant lower wages and the use of unqualified labour. In those times no one knew about the dangers created by increased energy use that the new machinery would bring.

Today, we understand the dangers that generation of electricity by nuclear power stations bring and although many of these can be managed reasonably effectively the great unknown danger is the storage of radioactive waste for thousands of years in the future. Past generations have given the present generations the problem of overcoming climate change. That legacy is going to be hard enough to deal with by future generations without given them another pos-

Sellafield

sible more insoluble problem of disposing of radioactive waste safely for hundreds of generations to come.

Nuclear generated electricity is considered to be carbon free, but this is a fallacy. The process, as we have seen, does not generate carbon dioxide, but the extraction of uranium does. In particular the extraction of low grade uranium is thought to produce as much carbon as the extraction and burning of natural gas. The storage of radioactive wastes requires construction of large concrete structures and this is a very carbon dioxide significant process. It is therefore wrong to characterise nuclear power stations as carbon free and whatever the reasons given for building them, reducing carbon dioxide emissions cannot be one of them.

Chapter 4

Energy's hidden bill

Energy, as we have seen, is central to our way of life and in many cases central to life itself. It comes at a price. When we most want a thing we pay a price and acquire the product. We know, for example in the case of cars, that they need servicing, filling with fuel and repairing from time to time. These costs are all visible when we make our purchase.

Energy is different because not only does our acquisition of it come at a visible cost, shown in our energy bills, our boiler servicing costs and even in the expense of sweeping chimneys, it also comes with a hidden bill. More often than not the purchasers of energy are not the same people as those who pay the hidden bill that use of energy purchase invariably presents.

These hidden costs result from pollution caused by energy used; climate changes costs caused by energy use and from energy raw material acquisition and processing. Pollution and environmental degradation are in themselves well understood but in many cases the costs associated with them are not considered costs for the energy account but costs for health or infrastructure.

It is worth looking at the energy costs that we all bear, albeit in different ways, as a result of the massive use of energy in industrialised countries.

Pollution created by energy is costly in obvious and also in subtle ways. Airborne pollution caused by smog in the 1950s led to many people being treated for lung diseases. It also meant higher cleaning bills for people, clothing and buildings. Some buildings were destroyed.

Acidic rain caused by airborne pollution damaged crops and killed fish. Livelihoods were in some cases ruined. There has been health costs associated with nuclear energy creation, when accidents happened. There are higher levels of breathing diseases especially asthma in young children which is claimed to be associated with pollution.

The by-product of energy use for transport results not only in atmospheric pollution, but also noise pollution. Vibrations damage property as anyone owning a building very close to a busy road knows.

Climate change is beginning to present "hidden" bills; ski resorts are experiencing shorter and shorter seasons; flood damage is destroying property. Unusual heat waves have recently killed people.

Now, on balance, energy provides far more benefits to people in the short term than it provides problems. The difficulty is to ensure that energy is used responsibly so that all users are aware of the hidden costs and arrange their behaviour to take account of these disadvantages. The right to use energy must come with a duty of equal weight to use energy responsibly and not to abuse its use so that its use does not present another human being with the hidden cost to pay, in the form of poor health, loss of resources or uncomfortable living conditions.

There is a complex relationship between life and climate that makes our world habitable. There is a very small range of temperatures that support life, compared with the total range of temperatures that we know of. Our own sun has a surface temperature of 6,000 degrees and an estimated core temperature of 15 million degrees Celsius. The coldest known temperature is -273 degrees Celsius which is one Kelvin, although in laboratory conditions colder temperatures have been achieved.

So the temperature of forty or so degrees within which human life (and most of animal and vegetation life) can be maintained is a tiny delicate fraction of all possible temperatures. When we do things that can potentially affect this complex relationship the result is very unlikely to be beneficial. The climate that supports life on earth does not need changing.

The way in which energy presents its bill is not regular or predictable. Our energy consumption is like a meal. Sometimes you have bought the food and cooked it yourself so after eating it there is nothing to pay. Sometimes you have to pay at the end of the meal. These days we are eating beyond our resources and living off credit that nature has provided. We will have to pay in the end perhaps years or centuries later, when the bill will be paid by other generations. Payment will be deducted or forcibly obtained; that is the way that nature works, but by a different person often in a different place who has no relationship with the people who ate the meal except the relationship of existing as a living being. There is no free lunch.

4.1 Energy and Atmospheric Pollution

As we began to use energy in increasingly large amounts, concerned voices began to be raised about the increased pollution. Some people pointed out that acid rain was poisoning our rivers and lakes. Others claimed that the use of fossil fuel could be linked to holes appearing at the North and South poles in the ozone layer of the atmosphere. Many eminent people warned that the globe might be becoming warmer and warmer and that climate change would cause irreversible damage to our environment and ultimately to ourselves.

Some of the fears and claims were wrong. It is now almost universally accepted, for example, that holes in the ozone layer are not linked to fossil fuel burning but to the use of certain types of refrigerants and propellants, chlorine and bromine compounds, which are produced by human activity.

What became obvious and what was eventually universally accepted was that burning fossil fuel causes pollution which is harmful to our health and there is a direct link to atmospheric pollution and fossil fuel burning.

The human being is a reasonably sturdy animal; he (or she) can live in both cold and warm climates. In addition, the human body can cope with a certain amount of impurities. Our nostrils filter dust before it reaches our lungs. We have many built-in features which enable us to live with some pollution but we cannot live well in a heavily polluted atmosphere.

In the period just after the Second World War, peacetime industry expanded to meet the needs of a world largely at peace. The increased production led to a great deal of coal burning and "pea-souper" fogs were commonplace.

Many heavily industrialised places have experienced these very thick fogs; London pea-soupers were infamous. It can be hard to imagine now, but these fogs were so thick that buses could only travel if the bus conductor walked in front of them, holding a burning torch to enable the bus driver to see the road. Walking in these conditions, even for a short distance, made one cough, splutter and gasp for air. These fogs regularly affected all large cities in the United

Kingdom (as well as many in places as far apart as the Po Valley and China) until the 1970s.

When the pea-souper came many people who suffered from breathing difficulties died; even healthy people were advised to stay at home. The very young could be damaged for life if they went out in a pea-souper. It could be and often was fatal for the very old and the infirm. Hundreds of people died and thousands became seriously ill each year as a result of these pea-soupers in the United Kingdom.

The 5th December 1952 is an infamous date in the history of London pea-soupers. The weather was considerably colder than usual and the sky was very clear. People were burning more coal than usual to keep warm. There was almost no wind.

By evening the fog had laid down all over the city. In central London the visibility remained below 500 meters continuously for 114 hours and below 50 meters continuously for 48 hours. At Heathrow airport visibility remained below 10 metres for almost 48 hours from the morning of 6 December. The smog remained for five days.

Although there were the inconveniences of no public transport with virtually all public events being closed, the most serious consequence was that thousands of people died. People first understood the large increase in mortality rates when undertakers ran out of coffins. Measurements at the National Gallery, in Trafalgar Square, showed 56 times more soot in the air than usual and seven times more sulphur.

The deaths which resulted from the smog were caused by pneumonia, bronchitis, tuberculosis and heart failure. Respiratory and cardiac distress meant that people were dying as they gasped for air; some died of asphyxiation in their sleep.

These days medical science understands that smog creates a short term decrease in breathing ability and an increase in chest pains; the lungs become inflamed and respiratory cells are permanently damaged. Asthma attacks increase due to higher levels of nitrogen dioxide.

The official figures reveal that this single pea souper smog killed about 4,000 people, but these figures only count those who died during the smog and for two weeks afterwards. However, as the smog happened near Christmas there were many Christmas deaths and the registration of these were delayed due to the holiday period. Deaths and delayed deaths means that it is estimated that you should add around 8,000 more deaths to the official figures, making a total of 12,000 deaths.

The way in which these pea-soupers – technically known as smogs – formed has never been completely understood but they are clearly directly linked to atmospheric pollution.

Smogs have great "optical" thickness which limits visibility to less than a few metres. What probably happens is this. Coal burning causes airborne particles of tiny dust. As the relative humidity grows, these particles take up water vapour around the nuclei of the dust and a haze is formed. When the relative humidity exceeds 100%, water starts to condense on these particles rapidly until their size has grown by a thousand times, and this forms the visible cloud known as smog. This is rather like an aerosol, which attaches a large amount of liquid to a small particle of solid.

Polluted air contains high concentrations of various water-soluble gases. The main ones are nitric acid, hydrochloric acid and sulphur dioxide. These gases are able to dissolve into the water-laden particles of dust before 100% relative humidity is exceeded. As the amount of dissolved gas in the aerosols grows, the aerosols are able to take up more water, and in this way a large fraction of the aerosols can to grow to cloud-droplet sizes without them condensing as rain.

These aerosols probably grow so large that they can block light efficiently and thus a pea souper or smog is created by the sheer volume of dense low lying clouds of vapour formed by particles of coal burning waste.

The United Kingdom introduced Clean Air Acts in 1956 and 1968 to prevent the smogs of the 1950s and 1960s. It was thought that these smogs were caused by the widespread burning of coal for home heating and for industrial processes. Local Authorities were empowered to control emissions of dark smoke, grit, dust and fumes from industrial premises and furnaces and to declare "smoke control areas" in which emissions of smoke from homes were prohibited.

Clean air legislation has been very successful; smoke control areas have been introduced in most of our large towns and cities in the UK and in large parts of the Midlands, North West, South Yorkshire, North East of England, Central and Southern Scotland. There have been substantial reductions in concentrations of smoke and associated levels of sulphur dioxide between the 1950s and the present day and pea soupers in the United Kingdom no longer occur.

However, while coal burning for domestic and industrial purposes has been significantly reduced year by year since the 1950s, the use of electricity has increased tremendously over the same period. The Clean Air Acts have been successful in reducing coal-created pollution and have led to people being healthier. As the Clean Air Acts succeeded in cleaning up the air in one respect so other intensive energy use, mainly of electricity, created other problems which while not as visibly dramatic as pea soupers and smog, are just as dangerous and deadly.

Although the Clean Air Acts did clean up soot from the atmosphere, in 1991 London experienced another great smog. Stagnant cold air settled over the city and trapped particles created by vehicular transport, created smog. Luckily, alert to the health hazards, the government warned people to stay at home. As a result "only" two hundred people died from the smog.

As we have seen, our main sources of energy are coal, oil and natural gas. A large part of these fuels is converted into the electricity that is our major form of power. The burning of these fossil fuels generates much of our electricity in the United Kingdom. The use of nuclear energy does not necessitate the burning of fossil fuels. A separate process creates heat. Virtually every time you use electricity – by making a phone call, washing your hands, turning on a light, watching television or making a cup of tea – you either burn fossil fuel or you create nuclear waste. Only a small portion of electricity is generated from benign non-polluting sources.

The basic principle that applies to all traditional power stations is that water is heated into steam by burning fossil fuels; the steam drives turbines, which generate electricity. By-products include steam, heat, gases and fuel waste. In order to generate electricity from coal, oil, gas or even nuclear fuel, the process involves heating water into steam. The steam passes through turbines, turning them, which in turn produces electric current. By this simple process current is created in the same way as current generated by a small bicycle dynamo, which can power a bike lamp, as long as the rider is turning it.

In addition to burning fuel to create electricity, we burn oil in the forms of petrol, diesel and domestic heating oil, which is usually kerosene and aviation fuel, which are also derived from oil. For the

New York City smothered in a fog caused by polluted air being trapped over the city by thermal inversion. Fortunately, this is still a relatively rare event.

past 15 years oil refineries have always produced more oil than the country demanded. The actual quantity of product refined has remained relatively unchanged from about 1986 onwards, although it fell by 29% between 1980 and 1985. We also burn natural gas for space heating, water heating and cooking.

When we generate electricity by burning fossil fuels, indeed whenever we burn fossil fuels, we create and distribute atmospheric pollutants, the main ones being blacksmoke, carbon dioxide (CO_2) and carbon monoxide, nitrogen oxides and sulphur dioxide (SO_2). I shall examine each of these pollutants that we have been, and still are, pouring into the air in increasing quantities.

Blacksmoke is suspended solid matter that is produced when fossil fuels, such as coal and oil, are burnt. Because the combustion of

Mexico City, one of the world's largest cities, covered in pollution. It is surrounded by high mountains so that thermal inversion trapping polluted air over the city is a common event.

these fuels is almost always incomplete, the unburnt solid matter is suspended in the air, soiling buildings and causing the haze which reduces visibility in fine weather. Most of the blacksmoke in our towns and cities comes from the combustion of diesel fuels. Particles of blacksmoke that are less than 10 microns in size have been linked to fatal respiratory diseases. The smaller the particle size, the more deeply it penetrates our lungs, often with terrible results.

Asthma sufferers will be very aware of the discomfort they have to endure caused by blacksmoke. A significant proportion of blacksmoke is caused by fossil fuels used in generating electricity, but most is caused by people's use of heat and transport directly. In

1993, 5% of blacksmoke was created by energy production, private households made 34% of it and transport 27%. If private households burning fossil fuels (coal and oil mainly) were to reduce their dependence on fossil fuels and used renewable sources, blacksmoke emitted in the UK would fall measurably and would significantly contribute to cleaner air and a healthier nation.

Carbon dioxide is created every time fossil fuel is burnt and every time living organisms breathe out. Plants use the carbon dioxide that all life forms exhale in the process called photosynthesis. Plants absorb carbon and emit oxygen. In this way plants clean the air and keep the carbon dioxide levels low.

Carbon dioxide is a long-lived pollutant, remaining in the atmosphere for between 50 and 200 years. It contributes significantly to the greenhouse effect, but it can be converted by photosynthesis in plants back to oxygen and plant material. Sadly, the equatorial rain forests, a great source of CO_2 absorption, are being rapidly destroyed.

It is thought that there have been natural fluctuations in CO_2 levels in the atmosphere. In the Holocene period, the past 11,000 years, CO_2 levels have varied enormously, according to studies from the Scripps Institute of Oceanography, San Diego, and the University of Bern. By taking core samples of Antarctic ice and analysing the air bubbles for CO_2 content, Scripps found that 11,000 years ago CO_2 was present at the rate of 286 parts per million by volume (ppmv) of air compared with about 190 ppmv 18,000 years ago and 285 ppmv in the late 1700s. The present rate is 381 ppmv. That is significantly the highest concentration that has been measured. This research supports the view that changes in CO_2 levels prompt changes in temperature, which in turn increase the CO_2 levels in the air. I shall deal with this later when I consider global warming.

> In the UK private households are responsible for the production of 22% of carbon dioxide (not to mention the production of 77% of carbon monoxide) according to government statistics.

Emissions of CO_2 from burning fossil fuels have actually decreased in the United Kingdom over the past 25 years. On the United Nations basis of measurement we emitted 19% less CO_2 in 1998 than we did in 1970 but 7% less in 1997 than we did in 1999. Emissions from transport, with engines burning more cleanly, fell by

39% between 1980 and 1998. Emissions from power stations, the largest single source, also fell greatly during this period but now appear to have levelled off. The United Nations calculates that during that time we discharged 149 million tonnes of carbon, in the form of carbon dioxide, into the atmosphere. However, others put the figure as high as 156 million tonnes.

If electricity is generated from natural gas instead of coal, roughly half the amount of CO_2 per unit of energy generated is produced. Nuclear-powered generators do not produce carbon dioxide, neither do hydroelectric generators. These forms of generation do have potential adverse environmental consequences however – either in the form of dangerous by-products in the case of nuclear energy, or damage to wildlife in the case of hydroelectric power.

Although, as I have explained, carbon dioxide is absorbed by plants, they cannot absorb all of the emissions that now take place. The surplus carbon dioxide in the atmosphere contributes greatly to global warming, as I shall explain later, so it is important to control and limit its emission.

Carbon dioxide is not the only carbon-based gas that pollutes as a result of energy being generated from fossil fuels.

Carbon monoxide is also created by imperfectly burning fossil fuels. This occurs not only in power stations but also in central heating and hot water boilers, coal fires and car exhausts. It is a lethal gas and is not recycled into oxygen by plants. If it is inhaled in significant quantities it kills. Households and cars create most of our carbon monoxide pollution.

In 2002, 3.2 million tonnes of carbon monoxide were emitted, a level 56% lower than 1990 and 63% lower than in 1970, so some positive progress is being made. Three-fifths of carbon monoxide emissions in the UK are thought to come from road transport, despite large reductions over the past thirty years due to tighter emission standards and the introduction of catalytic converters.

Nitrogen oxides are created when fossil fuels are burnt and some of the nitrogen in the air combines with oxygen. There is a range of chemicals involved but the most significant ones are nitrogen dioxide and nitrous oxide. High concentrations of nitrogen oxides damage health and they also harm plants. They also contribute to acid rain. Half of our production of nitrogen oxides comes from road transport with a further 20% from power stations. Burning coal creates twice the quantity of nitrogen oxides per unit of energy as oil,

Acid rain has virtually destroyed this forest at Blue Ridge Parkway in North Carolina, USA.

which in turn emits twice as much as gas. Although natural gas burns more cleanly, it still emits significant amounts.

The quantity of nitrogen oxides propelled into the air each year has been declining slowly but steadily since 1990, mainly due to catalytic converters on motor vehicles, and the reduction in the use of coal in generating electricity. It should not be thought, however, that the amount of these acid rain-forming chemicals in the air is being significantly reduced; we are simply poisoning ourselves more slowly than before.

The total level of nitrogen oxide emissions in 2002 (at 1.6 million tonnes of nitrogen dioxide equivalent) was 43% lower than in 1990, with substantial falls from both road transport and power stations, the two largest contributing sectors. Emissions from power stations have declined since the mid-1990s because of increased output from nuclear stations and due to combined cycle gas turbine stations replacing coal-fired generating power stations. In addition, low nitrous oxide burners have been installed at many coal fired power stations. Modern catalytic converters in vehicles have also reduced emissions.

Burning natural gas does not produce sulphur emission but coal and oils frequently contain sulphur impurities, so that when they are burnt sulphur dioxide is produced in the form of an acid gas.

Acid rain kills all the tress in a Czech forest near Litvinov apart from species resistant to acidity.

Sulphur dioxide harms people and damages life and buildings when deposited as acid rain. Acid rain describes the various acidic gases, mainly sulphuric, which when emitted by combustion into the atmosphere, can be deposited either in rainfall – as wet acid rain, or in dust – as dry acid rain.

Because of the wind, acid rain can be, and often is, deposited hundreds of miles away from where the pollution was created. It is estimated that the United Kingdom's emissions which created acid rain, are responsible for deforesting 16% of Norway's forests.

Lakes, streams and water courses have all been polluted with acid rain. Although some types of plants and animals can resist acid rain better than others, it does damage ecosystems.

Many older buildings were built from softer rocks that could be fashioned into interesting shapes – limestone is a good example. Others used exotic marbles. Both limestone and marble are very susceptible to attack by acid rain. Acid rain has caused damage to buildings such as York Minster, the Parthenon and the Taj Mahal.

In 2002, there were one million tonnes of sulphur dioxide emitted, 73% lower than1990 and 84% lower than in 1970. The decrease is probably due to lower coal and fuel oil consumption over

St Mary's Parish Church in Nottingham,
showing the stonework damage caused by
pollution.

the period, and the introduction of flue gas desulphurisation at two coal fired power stations, operational from 1994. Sulphur dioxide emissions from road transport have decreased by over 87% since 1998 following a reduction in the sulphur content of fuel.

Some sulphur dioxide is emitted into the atmosphere naturally, particularly in volcanically active areas. At Smoking Hills, in the northern extremes of the Canadian tundra, low grade coal at the surface sometimes spontaneously ignites causing sulphur clouds to be released. As the gas clouds move over the surface of the land, the soil and freshwater become acidified. Local deposits of metals are dissolved which releases poisons into the air. Places like Smoking Hills are fortunately rare in nature. Most acid rain is produced as a result of man's activities.

The pollutant gases that cause acid rain (sulphur dioxide and nitrogen oxides) also damage human health because they interact in the atmosphere to form fine sulphate and nitrate particles that are inhaled deep into the lungs. Fine particles can also penetrate indoors. Many scientific studies have identified a relationship between elevated levels of fine particles and increased illness and premature death from heart and lung disorders, such as asthma and bronchitis. While walking, bathing or drinking acid rain is harmless, breathing in the pollutants that cause acid rain is very harmful, and they are inhaled without the person noticing it.

Chemical smog under certain conditions some dramatically noticeable pollution can take place. One form of smog, known as industrial smog, is produced when sulphur dioxide and other pollutants are released by fossil fuel burning and are contained in a layer of cold air that is trapped by thermal inversion – a layer of warm air above it.

Smog can also be produced by nitrous oxides (usually due to the incomplete burning of petrol, a problem common to the internal combustion engines of all cars). This reacts with oxygen to form a complex mixture of pollutants – a photochemical smog, a noxious cocktail. Photochemical smog often combines with industrial smog to make an even richer, more lethal mixture.

There are many large cities associated with smogs or haze. Los Angeles famously had a yellow sky over it when viewed from a distance. This was the smog mostly caused by vehicles. Mexico City has a similar haze, caused not only by vehicles but also by the huge quantities of liquid petroleum burnt in the city for water heating.

Chemical smog over Los Angeles. This is mostly caused by vehicle usage.

> Energy production is responsible for a massive 44% of the acid rain (produced by both sulphur oxides and nitrous oxides); manufacturing industry and agriculture together account for 26%, and private households are responsible for 11%.

Atmospheric pollution is just one form of pollution that occurs when fossil fuel is burnt.

Pollution, in the various forms that I have described, is an almost invariable consequence of our generating power and using it. However, there are other harmful effects that are as insidious as atmospheric pollution and more dangerous. The most dangerous is global warming.

4.2 Energy and Global Warming

The theory of global warming is also called the Greenhouse Effect. Most people refer to its effects as climate change. It must be regarded as the most serious problem facing humanity today, more deadly and terrible in its effects than any known disease, human conflict or natural catastrophe. It is all the more frightening because it neither kills nor ravages like an epidemic, but strangles life out of the planet slowly but surely. It causes the very force that makes life possible on earth – the sun – to become the force that will ultimately destroy that life.

Everything originates with the sun, which until now has benefited our planet and created the conditions for life. The sun emits a huge amount of energy in the form of electromagnetic radiation, which has a very short wave-length. About 30% of the sun's radiation that reaches the earth is reflected straight back into space, because short wave radiation is harder for the atmosphere to absorb than long wave radiation. The remaining 70% penetrates the atmosphere. Some is retained there while the rest reaches and warms the surface of our planet and its waters. This warming drives the water cycle we are all familiar with – it causes water to evaporate and form clouds which then precipitate rain hail and snow.

Some of the radiation absorbed at the surface is re-radiated in the form of long wave radiation which the atmosphere absorbs more easily than short wave.

Physicists have been pondering the Greenhouse Effect for far longer than most people would imagine. In December 1895 Svante Arrhenius presented a paper to the Royal Swedish Academy of Sciences about, as he termed it, "The Influence of Carbonic Acid in the Air upon the Temperature of the Ground". Apart from his own observations, which concentrated on what we would now call acid rain, Arrhenius drew on observations by the Irishman, Tyndall, in 1865, and the French scientist Fourier in 1827, to the effect that the atmosphere operates like "the glass in a hot house", trapping certain rays and letting through others.

Arrhenius concluded that increased amounts of carbon dioxide in the air made the atmosphere warmer. He believed that geological fluctuations in carbonic acids mainly from volcanoes, rock weathering and decomposition, were a cause of the Ice Ages. He blamed the increased amounts of carbonic acids that he was able to establish by his methodology in 1895 on coal production and use.

The modern theory of the Greenhouse Effect adds to these pioneering ideas. In addition to the trapping of heat by the atmosphere, greenhouse gases (mainly water vapour and carbon dioxide), also trap some of the outgoing energy before it can escape. As our way of life means that increasing amounts of carbon dioxide in particular are being emitted, the planet is warming up at a far greater rate than before, some would say at a dangerous and life-threatening rate.

Of course, there are many opinions about the degree of danger

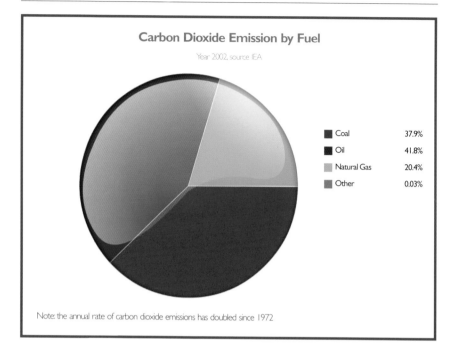

Note: the annual rate of carbon dioxide emissions has doubled since 1972

The different amounts of emissions of carbon dioxide according to the fuel consume in energy production. The rate of carbon dioxide emissions has doubled since 1972. Carbon usually takes about 100 years to break down in the atmosphere.

that we face from global warming. We should treat them with some caution and rely on those we perceive to be the most authoritative.

The United States Environmental Protection Agency, for example, believes that since the beginning of the Industrial Revolution carbon dioxide concentrations have increased by 30%, and those of nitrous oxides by 15% and those of methane more than doubled, and that these increases significantly enhance the ability of the earth's atmosphere to retain heat. They cite the following in support of their views:

The 10 warmest years of the 20th Century occurred between 1984 and 1999; 1998 was the warmest year. Snow cover and floating arctic ice have decreased. Global sea levels have risen by 10 to 25 centimetres over the past 100 years. Glaciers are in the main melting. Precipitation over land has increased by 1% and the frequency of rainfall in the United States has also increased.

If the United States Environmental Protection Agency is right then

The Perito Moreno glacier in Patagonia, Argentina, which is shown in this picture rupturing. Most glaciers (but not all) are melting as the climate changes.

there is indeed cause for grave concern. All commentators and scientists agree that water vapour and carbon dioxide trap heat in the atmosphere. If the temperature rises – even by, say, 1° Celsius over the next 100 years – the soil will become warmer and less moist, vaporation will increase causing more water vapour to exist in the atmosphere, and this will cause the atmosphere to retain more heat.

It is as simple as that.

Are temperatures rising? Of course as far as we can tell the climate of this planet has always been changing. The evidence of this is in the signs of glacial movements in England which have left behind debris and shaped the countryside.

Some observations over the past 100 years indicate a rise in that period of up to 1° Celsius. Global warming will cause rises in sea levels (mostly caused by the expansion of water as it becomes warmer with help from melting ice caps). Warmer water can retain less oxygen than colder water and gives rise to more rainfall in higher latitudes and less at the equator. This is because precipitation is caused by warmer, moisture-laden air meeting colder air; if the colder air moves further north so will the rainfall. This will cause deserts to expand further still; they are already expanding as a result of injudicious farming and forestry methods.

It is expected that apart from all the climatic change, there will be more violent rainstorms and drier soils with huge regional variations.

It should be pointed out that there is a division among informed scientific opinion about whether global warming is occurring and if so, whether its causes lie in the sheer volume of fossil fuels the world is now burning. Some scientists believe that the globe is not actually getting warmer. Sceptics argue that these are the scientists working for oil and coal businesses, but this is a disservice to an informed and honourable body of scientific opinion.

Accu-Weather, the leading commercial forecaster, reports that land based weather stations show an increase in temperature of about 0.45° Celsius over the past century. This, of course, may be no more than a normal climatic variation.

Weather satellite data appears to indicate a slight cooling of the climate over the past 20 years. The satellite data can be very confusing, especially the satellite measurements of the part of the earth's atmosphere which covers the area from the surface of the earth to a point some 10 to 15 kilometres high – the lower troposphere.

Some sources claim that 98% of the greenhouse gases are natural – mostly in the form of water vapour. Therefore, they argue, only 2% of the greenhouse gases are man-made – a tiny fraction of the amount involved. This seems to ignore the possibility that even a small amount of man-made greenhouse gas will marginally heat the globe causing more water vapour to be produced naturally.

Others say that the satellite data contains errors; when these are corrected, they will prove that the globe is getting hotter. This has been vigorously denied by the fossil fuel industry, although less so now. The argument centres on the accuracy of the instruments used to gauge the microwave radiation given out by oxygen molecules in the atmosphere, from which the air temperature can be calculated. Radiation levels are measured at a range of altitudes, including in the lower troposphere, so that the way the temperature measured is isolated is necessarily complex and the precise position is not certain.

The statistical complexities can be used by supporters of both

Flooding in Bihar, India. Heavy monsoon rains submerge villages, roads and fields. The flood waters also come from the Bagmati river branch of the Ganges River in Muzaffarpur District.

sides, yet they are crucial both to the future of the multi-billion pound fossil fuel business, and even more importantly to the future of our planet. Whatever the precise position may be, the vast majority of scientific opinion agrees that global warming is taking place.

Some scientists take the view that increased greenhouse gases are insignificant factors in global warming. Dr Sallie Baliunas of the Harvard-Smithsonian Center for Astrophysics believes that the real cause of global warming is the fluctuation that takes place in solar radiation. The sun does not emit its radiation at a steady rate but variably. Sunspots and unexplained phenomena can also vary the amount of heat radiated and thus affect the temperature of the planet.

It is highly significant that best placed climate experts generally agree that global warming is taking place. They also point to the fact that glaciers in the Alps are melting, the sea level has risen by 10 to 25 cm over the past 100 years and sea corals are being bleached by high sea-surface temperatures. These factors are all consistent with an increase in global air temperatures.

Extensive studies of the United States' climate show that night time temperatures have generally increased more than daytime tem-

The aftermath of Hurricane Katrina showing New Orleans on 3rd September 2005.
Opposite: *The scene at I-10 and Causeway Boulevard after Hurricane Katrina.*

peratures. Extreme climatic events, such as tornadoes, hurricanes and storms have measurably increased in the United States, and it seems from our personal experience that these extreme events have also increased in the United Kingdom. In October 2000 parts of Bognor Regis, in Sussex, were devastated by two tornadoes occurring within days of each other. Heavy rainfall events in the US have heightened in intensity.

On 28 August 2005 Hurricane Katrina gained strength as it travelled across the Gulf of Mexico. It was estimated to be one of the most violent storms ever in that region. New Orleans, in its path, was ordered to be evacuated but there were still thousands of people in the city, mostly below sea level, the hurricane struck on 29th

August. The city's flood defences were breached and by 9th September 2005 80% of New Orleans was under water.

Previous hurricanes in the United States cost hundreds of lives. Katrina cost thousands. Although some were blaming specific authorities and institutions over the handling of the disaster, it seems that there was a fundamental failure by the United States in its attitude to climate change. There seems to be a failure to recognise that the weather is getting more extreme and we have to plan for these new levels of extremes.

Extreme weather planning has to go hand in hand with not only a systematic programme for carbon reduction but also strengthening of sea defences in low lying areas as well as recognising that these disasters will almost certainly become more and more frequent. The weather is more powerful than any world leader.

Almost all specialists agree that without drastic steps to curb greenhouse gas emissions, the average global temperature will increase 1 to 3.5° Celsius during the next hundred years because effective levels of carbon dioxide are expected to double, probably within the next fifty years.

The human race is adapted to live within a very small temperature range. We cannot easily live at the poles or in the middle of a waterless desert. Even a change of 1° Celsius would greatly affect our way of life and probably the ability of many people to survive. During the so-called Little Ice Age, a period lasting from 1500 to 1850 when there were extensive glacial advances, the global temperature was only about 0.5° Celsius lower than it was in 1900. The Little Ice Age had a profound effect upon agriculture and was even responsible for migratory movements of humans leading to conflict and wars.

It is not surprising, therefore, that there is a general consensus amongst scientists that the higher temperatures projected for the next century will cause more frequent and intense heat waves, wide-scale ecological disruptions, a decline of agricultural production in the tropics and subtropics and the continued acceleration of sea-level rise.

Acceleration of warming is to be feared; deep beneath the floor of the seas there are large deposits of methane hydrates. These are stable and harmless at present. Temperature rises might well cause the hydrates to release methane gas. Methane is one of the most powerful greenhouse gases, some 21 times more powerful in the global warming contributories than carbon dioxide.

> One broad area of agreement is that levels of greenhouse gases in the atmosphere, primarily carbon dioxide, methane, nitrous oxide, ozone and hydrocarbons have grown significantly since pre-industrial times. During this period, the carbon dioxide level has risen 31% to more than 381 parts per million, methane 145% to more than 1,700 parts per billion and nitrous oxide 15% to more than 300 parts per billion.

The banning of CFCs helped to prevent the increase in one type of greenhouse gas, the carbon-hydro-fluorides, and probably put us back on the path towards mending the ozone hole, but in reality nothing has been done to prevent the continued increases of carbon dioxide in the atmosphere.

There has been a lot of speculation recently about whether more

381 ppm and rising. This station, situate at the Zugspitze, Germany's highest mountain, measures carbon concentration in the atmosphere. You can see the expanse of Genersys thermal solar panels at the front of the building, which is also served by a ground source heat pump. (ThermoSolar A.G.)

In the Fray Jorge National Park, Chile, the desert bursts into leaf after El Niño rains.

frequent hurricanes and more intense and longer lasting El Niño conditions are related to global warming. A warmer sea surface is the primary feature of global warming that might cause more significant hurricanes, but ocean circulation changes may counter the effects of this added warmth.

Since 1976, El Niño episodes have been stronger, more frequent, and more persistent than they were earlier in the century. Some of the models suggest that stronger, more frequent El Niño episodes would be a tendency in a warmer world. But there is no consensus in the scientific community about how these would change. Scientists are doing a lot of research to see if they can establish a cause-effect relationship between global warming and the change in the pattern of El Niño phenomena. In the western Caribbean off the coast of Belize, around the Cook Islands, and in the Philippines, massive coral bleaching has been observed since 1983. Coral reef bleaching results from the expulsion of symbiotic zooxanthellae algae from the coral reefs. The algae provide reefs with most of their colour, carbon, and ability to deposit limestone.

Raymond L. Hayes, Professor of Anatomy at Howard University, Washington, D.C., has studied the bleaching of corals. He found that coral bleaching occurs when the ocean temperature exceeds 30° Celsius for more than two weeks. The normal maximum ocean temperature in regions where corals live is around 28° to 29° Celsius. If the maximum temperature rises a modest one degree Celsius, reports Professor Hayes, the coral becomes bleached and cannot take in enough food or oxygen. If water temperatures return to normal the following year, coral reefs may recover, but if the coral is subjected to continued temperature rises, the reefs may die.

Another important indicator of a warmer planet is the retreat of alpine glaciers. In the tropics, every glacier appears to be retreating and the rate of retreat is accelerating. In Venezuela, for example, three glaciers have completely disappeared since 1972. In the temperate zones, most of the glaciers are also retreating.

Peru is particularly concerned about the accelerating melting of glaciers in the Cordera Blanca region of the Andes. The glaciers there provide Peru with irrigation water for its coastal desert and feed rivers that are dammed for hydroelectric power. Loss of glaciers would harm both power production and agriculture for the country.

In 2005 one Swiss ski resort covered part of the Gurschen glacier with reflective material to reduce the ice melt and delay retreat

*These corals, off the Maldives in the Indian Ocean, are bleached due to high
ocean temperatures.*

of the glacier. The Gurschen glacier has sunk 20m (66 feet) in the
last 15 years, making the resort of Andermatt's ski slopes almost
inaccessible. Without the glacier, the resort's raison d'etre will cease.
Professor Wilfried Haeberli, of the University of Zurich, believes
that 70% of Switzerland's glaciers will disappear in the next 30
years, due to the effects of global warming.

It was thought that climatic changes in the past happened very
slowly. However by measuring oxygen isotopes and dust in Green-
land's ice cores and by studying glaciers in the Andes and in New
Zealand's Alps, scientists have found that some of these temperature
changes were very sudden, sometimes fluctuating 5° to 10° Celsius
in less than twenty years.

Can the world adapt to global warming? If the climate does
change can the world's natural ecosystems simply adapt to the
changes with little adverse effect upon them? I doubt it.

Although we can point to the known climate changes that ecosystems have adapted to – the Ice Ages for example – I do not believe
that the ecosystems can adapt satisfactorily in the sense of preserv-
ing the range of life that they now maintain.

The main reason is that the previous climate changes took place

These researchers are a examining the ice core in Northern Alaska for evidence of global warming and ozone depletion.

over thousands of years, whereas global warming is likely to cause significant climate change over the next ten or twenty years. Life can often adapt to very slow climate changes. There is no evidence of life being capable of adapting to climate change at the speed that global warming will bring about. It is theoretically possible that life will be capable of evolving and adapting, but natural history teaches that not all life will be capable of adapting and there is no assurance that human life will be able to adapt quickly enough to survive.

> One notable thinker, Professor Stephen Hawking, expects that humans can only adapt quickly enough to overcome its threatened extinction as a result of global warming if they can migrate to new worlds.

In addition, previous climate changes caused alterations in the natural environment. As the ice caps gradually retreated more suitable vegetation took root and animals slowly migrated or adapted. Today humankind has vastly altered the natural environment. Tropical rain forests have been cut down; temperate woodlands have been ploughed up. Pollution has also caused changes. These factors have probably made the ecosystems less robust and more dependent upon human management. They are now fragile and likely to be grossly affected by climate change.

Studies have shown that life on earth has always moved to favourable locations. Species of all life forms have been able to shift slowly into areas where they can prosper most. For example, studies of plant fossils appear to indicate that plants can migrate to more favourable areas at rates of between 40 metres a year for the slowest and 2 kilometres a year for the fastest plants. However other studies indicate that temperatures could increase at a speed that require plants to migrate at rates of between 1.5 and 5 kilometres a year.

If plants cannot move as quickly as they need to, then individual plant species will die out and large volumes of vegetation will be lost. The globe's capacity for converting carbon dioxide into oxygen would be reduced. This could in itself contribute to further and even faster global warming.

I have so far considered the effect of carbon as a climate changer but I should also consider methane's role in climate change.

Methane As a trace gas in the atmosphere, methane has a more warming effect than carbon dioxide. It is about 21 times more on a molecule-by- molecule basis, and 58 times more on a kilogram-for-kilogram basis according to the Intergovernmental Panel on Climate Change.

Concentrations of methane in the atmosphere have more than doubled in the last two hundred years. This is very much greater than the increase of carbon in the atmosphere.

The good news is that methane has a relatively short life span in the atmosphere before it combines with hydroxyl radical to form water. The life span is some eleven years but in that time it can do much damage.

Some methane is naturally occurring and is vented through volcanoes and other breaks and fissures in the crust of the earth. The amount of methane naturally released has never been measured but scientists assume that as in most natural regimes the amount of naturally introduced methane is probably balanced by the methane that has decayed naturally over the same period.

Much methane is introduced into the atmosphere by anaerobic bacteria that in the absence of oxygen form methane. It is these anaerobic bacteria that decompose vegetation and animal refuse in wetlands and marshes. This methane is often locally known as swamp gas, or marsh gas. Methane produced in this way accounts for about a fifth of the methane produced, around 115 million tonnes.

Another important source of methane is gas from animal intestines. Here cattle are the chief producers but other grazing livestock, such as sheep, also produce methane. Anaerobic micro-organisms in the intestines of animals make digestion of vegetation possible – but the methane the micro-organisms produce is discharged into the atmosphere. This so called "enteric fermentation" together with agricultural production (such as rice paddies) contributes around 140 million metric tonnes of methane to the atmosphere each year.

Other anthropogenic methane sources include the extraction of fossil fuels (estimated at around 100 million metric tonnes per year) landfills (25 million metric tonnes per year) and biomass burning (40 million metric tonnes per year).

These estimates have changed significantly over the past few years and there is no real reliable data to be sure of even roughly how much methane is being introduced into the atmosphere although all the data indicates increasing amounts on a year by year basis.

Whatever the uncertainties, it does seem probable that much of the observed increase in the concentration of atmospheric methane since the beginning of the industrial revolution has been caused by human activity. It follows that, without technological change, further increases in human population, industry and agriculture will bring further increases in atmospheric methane concentrations.

Climate change, caused by humans introducing more methane and carbon into the atmosphere is a serious problem. It will be possible to fill many books with differing projections by scientists and thinkers. Some believe that sea levels will drastically rise; others believe that they will only marginally rise. Some predict storms tempests and hurricanes. Others envisage the South of England having the same climate as Greece. Others foresee deserts where now lie fertile fields.

The work done by the Tyndall Centre for Climate Change Research and Environment Agency tries to project what will happen to the climate in the year 3000. The next 25 years are crucial, their research indicates. Among the scenarios projected by the study very high temperature rises – as much as 15° Celsius, sea levels rising by 11.4 metres, rapid and abrupt climatic changes occurring, even long after emissions cease and the ocean pH falling, causing a dramatic loss of plankton and ocean life. Of course if any one of

Scientists in New Zealand have clothed this sheep with a device so that they can measure the methane it produces. The sheep seems quite unaffected by its garb.

these things happens it spells disaster and massive loss of life as well as a complete change of life as we have known it over recorded time.

Guessing the future is never likely to be profitable, because in the end the sheer variety of guesses can dilute the seriousness of the exercise. It is probably better to consider this; if climate change has only a remote chance of adversely affecting the future of a planet – let us say less than 2% - dare we taken even that small chance?

If a small child had a 2% chance of being killed when crossing a road by itself every parent would surely not let the child take the chance. We can be no less careful with the life of our planet than we are with the lives of our children.

4.3 Energy and Global Dimming

Compared with the theory of Global Warming, the concept of Global Dimming is new and relatively unknown. Climatologists have been studying the warming of the earth but from time to time data was discovered which seemed to indicate that notwithstanding the warming of the earth the sun seemed to be measurably less powerful. Measuring the heat of the sun when it strikes the earth is different from measuring the heat of the sun in space. The sun's heat strikes the planet as light and is measured as irradiation or insolation.

In the1990s, scientists found that light levels were falling significantly. Studies showed that sunshine levels in Ireland, in the Arctic, the Antarctic and in Japan were falling.

Most scientists assumed that these measurements were simply wrong; they were inconsistent with the measured increase in heat in these places. All of these places were experiencing hotter climates, which most scientists attributed to global warming or the greenhouse effect.

Scientific opinion began to change when in 2001 scientists at the Volcani Centre in Bet Dagan, Israel published data that tended to show that, on average, the amount of solar radiation reaching the Earth's surface reduced by between 0.23 and 0.32% each year from 1958 to 1992.

Techniques for measuring solar radiation forty years ago were suspect, many scientists thought. Many of them discounted the earlier measurements. More data was collected and was thoroughly analysed. Now many leading climatologists believe the data indicates that that the solar irradiance at the Earth's surface has decreased, notwithstanding the fact that the climate is getting hotter.

The best way to picture this is to imagine a field where you measure the air temperature and also the amount of light hitting the floor. If you find the temperature is constantly rising but the light hitting the field is constantly decreasing you will understand the effect. You know that light converts into heat when it strikes a body. Heat and light are both the same form of energy. Your problem is to discover why there is more energy on the field when there should be less, because there is less light.

Scientists found this phenomenon. In particular they discovered that the radiation that has measurably decreased is in the spectrum of visible light and infra-red. The ultraviolet light is not dimming, but there is less infra-red radiation and less visible light.

If light is getting lower in intensity when it reaches the earth, then we have to look for reasons why this should be the case. The most viable theory is that air pollution is responsible for global dimming. As we have seen, energy created by fossil fuel always creates by-products of dust, soot and particles which reflect sunlight. It bounces off the particles and diffuses. These particles also appear to aid for-

Mexico City in its customary haze which significantly limits the insolation the city receives.

mation of bigger, longer lasting clouds, which also prevent more light reaching the surface.

Clouds absorb heat from the sun and they also absorb heat radiated back from Earth. During the day the absorption of sunlight cools but at night the clouds reflect and absorb re-radiated heat from the Earth and this warms us.

The increase in particles of fine dust then seems to dim light reaching the planet while helping form clouds that retain heat and reflect heat back to the surface of the earth. In this way global warming occurs at the same time as global dimming.

If global dimming does exist, what does it mean for us? Temperate zones, like North West Europe, will be affected in a number of ways. Crops will grow in marginally less abundance; there will be more cloudy days and less light but more heat. However, equatorial areas may be actually experiencing the effect of global dimming in a very alarming way.

Infra-red and visible light intensities remain the same, so we will still be affected by sunburn, which is caused by excessive exposure to ultraviolet radiation while enjoying less light.

One study measured solar radiation attenuation in Mexico City and compared it with the solar radiation 8 kilometres north east of the airport. It found that Mexico City receives on clear days during the dry season 21.6% less solar energy than the rural surroundings. Mexico City is surrounded by mountains. Its 20 million inhabitants drive cars and generally source their heat for space and water from liquid petroleum gas, which burns in fairly unsophisticated boilers discharging dust to mingle with the dust caused by the city's traffic.

The mountains add a thermal inversion effect, so that the dust is trapped over the city. The large difference in solar radiation between Mexico City and a location only 8 kilometres away tends to show that there is a dimming of light caused by these pollutants.

If global dimming exists, there are several potential consequences, some of which are quite frightening. There is the effect on food supplies. People in many parts of the world depend on seasonal rains. The monsoon is a good example. It enables people to grow the crops that they need. These rains are caused by the heat of the sun. It is possible that there is a direct link between the dimming of the globe and the failure of monsoons in Ethiopia, which led to droughts in which millions perished.

> Professor Veerabhadran Ramanathan of the University of California believes that global dimming could lead to drought affecting areas like India, Bangladesh and South East Asia. In India the monsoon has been coming later and later in recent years. Ramanathan postulates that one year it may not come at all. If the monsoon does not come the climatic "knock on" effect will be unpredictable but it is bound to be dramatic and malignant.

It is thought that vapour trails of aircraft are a significant factor to global dimming because they not only actively contribute to particulates in the atmosphere (probably around 3% of them are caused by aircraft) but also they deliver the particulates high in the sky where they can affect light the most.

Vapour trails from a jet aircraft There are thousand of aircraft journeys every day in the United States alone, each one burning volumes of fossil fuel and leaving the waste in the skies. After the deplorable terrorist attack on New York's Twin Towers on the 11th September 2001, the United States Government stopped all commercial aircraft flights for three days.

Climatic measurements taken in North America during the three days when commercial aircraft were not operating showed an unusual effect. Temperature differentials actually increased between the lowest and the highest by more than one degree Celsius during the three days when commercial flights were grounded. In other words, there was measurably more heat when the aircraft were stopped from flying.

This means, it is argued, that global dimming is actually slowing up the process of global warming. It is possible that if all particulate discharge were stopped we would find temperatures rising more dramatically than they are at present. If we reduce particulates by burning leaner and cleaner fuels we could well be accelerating global warming. By trying to solve the problem of global warming we could be simply accelerating it, if we deal with it piecemeal.

Climate is sensitive and complicated. It is probably too complicated for us to model with any degree of accuracy. But some modelling by reputable scientists indicates that the global dimming and global warming together create far more serious problems than we can imagine or possibly solve.

4.4 Energy & Carbon Capture

Humankind is ingenious. We see that by creating energy from fossil fuel we are emitting carbon into the atmosphere. If we assume that it is not in our long term interest (as a species) to emit large quantities of carbon into the atmosphere, then we only have three options. First, we can stop using fossil fuel. This is such a drastic remedy that no one will want to do this. We depend on fossil fuel for our wealth and comfort, and indirectly for our very existence. We need it but without its bad side effects.

Secondly, we can look for ways of creating energy without creating carbon dioxide as a by-product. There are ways of doing this, and many of them are mature and viable. We shall discuss those ways in more detail later.

Thirdly, we can continue to burn fossil fuel, create the carbon but capture it at source and store it, so that it is not emitted into the air and cannot affect our climate. This third possibility, called "carbon capture" or carbon sequestration, has been taxing the ingenuity of the minds of humans.

Capturing the carbon and storing it in geological formations is actively being proposed by many organisations, particularly those industries closely associated with fossil fuel burning. If carbon can be effectively captured and stored it might well prove to be the panacea that humanity appears to need.

The main businesses researching into this field are also those closely connected with the fossil fuel industry. The leading lights include BP, Exxon, Shell, EdF, Ford, Statoil, Texaco, Total Fina Elf, RWE, and General Motors. Research is carried out at universities with the help of significant sponsorship from these and other companies.

A leading researcher is the IEA Greenhouse Gas R&D Program, which is supported by 19 countries and the European Union with sponsorship from many oil and energy companies.

IEA is actively assessing the storage of carbon in deep underground saline aquifers, under the North Sea, called the Sleipner project, because the storage is the above the site of the Sleipner field from which large quantities of natural gas has been extracted. The project is being funded by Statoil, at an estimated cost of 350 million euros.

The Sleipner field produces natural gas which contains around 9% carbon dioxide. The market requires the carbon dioxide content to be reduced to around 2%. At the present rate of extraction of natural gas from the field the carbon dioxide that must be removed from the gas is about a million tonnes a year. The normal way for a gas extracting company to deal with the carbon is simply release it into the atmosphere. That way the gas company can deliver natural gas lower in carbon dioxide. Statoil is a Norwegian company operating in the Norwegian North Sea and Norway imposes high taxes on carbon emissions.

Natural gas production from the field started in 1996, with the CO_2 produced being injected into the Utsira aquifer formation, some 800 metres below the bed of the North Sea. Had this process not been adopted, and the carbon dioxide produced been allowed to escape to the atmosphere, the licensees of the Sleipner would have had to pay over a million Norwegian Kroner a day in Norwegian taxes roughly equivalent to £110,000 a day.

Statoil's strategy of capturing the carbon dioxide could not be said, then, to be wholly altruistic but nevertheless it is worth understanding what they are doing. The carbon dioxide removal process used to remove CO_2 from the high pressure natural gas stream is based on amine scrubbing technology. The amine solution collects the carbon dioxide from the gas. Subsequently, the collected carbon dioxide is removed, mainly by flash regeneration, and the amine is re-circulated for use in collecting more carbon.

Energy released by the amine treatment process runs two generators, yielding 6 MW of power, which is used on the gas platform. The carbon dioxide is then injected into the Utsira aquifer. A small amount of natural gas is mixed with it to improve its overall condensate. The Utsira Formation is sandstone between 200-250metres thick. The carbon dioxide is injected and stored into the Utsira formation 1,000 metres beneath the seabed. One million tonnes of carbon dioxide are stored in the sub-surface each year and the field operators think that the Utsira Formation is capable of storing up to 600 billion tonnes of carbon dioxide.

This is, of course, carbon capture in a very specific process. The underground storage chambers in the sandstone have been seismically mapped and can be monitored for leakage.

Of course, if a safe way of capturing and storing carbon can be found then these research bodies will have done a great service to humanity, but there are some real problems that will have to be overcome. It is not really known whether the seals and characters of saline aquifers and sandstone make for a really permanent storage facility. A large scale leak from a carbon storage facility in a hundred years from now could cause great and immediate harm to the planet as well as rendering the whole activity pointless.

The health effects of these processes are not known. No one really understands what would happen if there was slow leakage of carbon dioxide through soils and pipelines. It is very difficult to predict this with accuracy.

Nature's way of storing carbon safely was to store it in living matter, which when it died held huge quantities safely. Coal, oil and natural gas deposits are such natural stores and the artificiality of storing carbon dioxide in aquifers is such that great care must be taken before this becomes widespread.

Chapter 5

The renewable technologies

A school with a thermal solar system. (Ecowarm)

We have seen that the traditional fossil fuels of oil, coal and gas provide us with major sources of energy. Coal was the first energy source that transformed Europe and North America and laid the foundation of the Industrial Revolution over two hundred years ago. At the beginning of the last century oil became important and the development of the internal combustion engine created greater and greater demands for oil until it became central to our energy requirements.

Oil is still essential today and as a result huge power and control is vested in a few multinational corporations and some relatively small oil-producing states. Gas became important in the United Kingdom when large gas fields were found under the North Sea.

Lastly nuclear power became important in the middle and later parts of the last century.

The new century must be the century of renewable energy or it may end in disaster for our race. Professor Stephen Hawking has warned that unless humankind colonises planets in outer space, it will have no future, due to global warming. The danger of global warming may be that it is simply irreversible, or more likely, that humankind will not heed the warnings and the dangers and will do nothing to reverse the burning of fossil fuels and other malpractices which are leading to a crisis that will affect the whole planet.

I believe that it is undesirable for energy to be controlled by multi-national corporations or by sovereign states, whether large or small. Energy is so important, in that it is needed for our health, comfort, employment, and security and well-being, that it should be diversely controlled and capable of being obtained from as many different sources as possible. There is a good arguable case for national energy supplies to be controlled by the state they serve.

Government policy in the United Kingdom has always sought to achieve diversity of supply. With current developments it will, in the near future, be possible for the United Kingdom to be supplied with energy from many diverse sources, none of which could be easily controlled or threatened by monopolies or other sovereign states but the government must have the vision to plan for this and encourage this.

The United Kingdom has concentrated its endeavours so far on developing renewables mainly by the use of wind turbines to create electricity. The reasoning was that electricity generation by fossil fuel is more environmentally damaging than heat generation by fossil fuel. I am not sure that this reasoning is entirely correct but it did lead to a policy of working to generate grid electricity from renewable sources, ploughing much taxpayers' money into this. However, renewable sources of electricity generation in the United Kingdom account for only a tiny proportion of the total amount of electrical generating capacity.

By 1999 electrical capacity from renewable sources in the United Kingdom had increased to 3 gigawatt. Whereas the other sources of electrical generation typically use 50% of capacity, the average capacity use of renewables is about 80%.

The figures include so-called "bio-renewables" such as wood, straw and other fossil based fuels, which can involve pollution. They also include free sources such as wind, waves and the sun.

It is here that we can discern the beginnings of what we hope will be the trend for the foreseeable future, which is, the generation of all energy, not just electricity by using the wind and the producer of wind power – the limitless power of the sun.

There are ten environmentally friendly ways of creating energy, which will over the next few years prove as significant and as important to humanity as the traditional forms of energy production. These are: thermal solar panels; fuel cells; photovoltaic cells; wind turbines; tidal power; wave power; ocean thermal power; geothermal heat pumps; air source heat pumps and biomass. Some are more environmentally friendly than others, but all are more environmentally friendly than fossil fuel.

Although hydrogen might have limited uses as a source of energy, it has great potential as a carrier of energy, so I shall discuss hydrogen and its future role in our lives in this section when I describe fuel cells.

In 2001 there was very little public information about these renewable sources, particularly in the United Kingdom. Today, a few short years later, information about them is more widely available and the Government provides some explanation of them in White Papers and consultation documents. I think, nevertheless, that the simple explanations are often unintentionally misleading in that they

A Genersys thermal solar system built into a roof, using horizontal panels. (Cel-F Solar)

do not provide detailed information which can give an enquiring reader a good understanding of what actually happens in these renewable sources, what the advantages, disadvantages and limits are, and perhaps most importantly the science on which these energy sources are founded. With that understanding I think that individuals will be better placed to choose which methods of generating renewable energy are best for their own homes.

It is possible to combine these individual technologies so that, for example, solar collectors and photovoltaic cells can be used in complementary ways. Indeed it is desirable so to do, in my view, and so the last part of this section seeks to explain this important point.

5.1 Photovoltaic Cells

The French physicist, Edmund Becquerel was literally the great grandfather of electrochemistry, in that he was the first of four generations of French scientists who each made significant contributions spanning two centuries. He discovered the photovoltaic effect in 1839. Becquerel observed, while experimenting with an electrolytic cell made up of two platinum electrodes placed in an electricity-conducting solution that electric current generation increased when exposed to light. He could not explain why.

In 1873, Willoughby Smith discovered the photoconductivity of selenium (an element derived from copper) and, four years later, the photovoltaic effect in solid selenium was noted by Charles Fritts, an American inventor.

Fritts created selenium squares, which he attached to a brass plate. He covered the squares with a transparent gold film and noted that small electrical currents were produced. However, less than one tenth of 1% of the light was converted into electricity, so there was no apparent benefit in the discovery.

It seemed at the time that these eminent scientists were observing and experimenting with no obvious practical end. But often the trial and error of apparently meaningless experimentation and observation leads to discoveries that enhance the quality of life and benefit mankind.

In the 1870s another team of scientists, led by Heinrich Hertz, developed photovoltaic cells made from solid selenium. These could convert light into electricity at an efficiency rate of nearly 2%. In 1905, Albert Einstein published his paper explaining the photoelectric effect. This was based on work by Max Planck, a fellow physicist. His theory was that energy could only be released or absorbed by atoms as atomic radiation in packets called quanta. Einstein used this idea to show why light above certain frequencies shining on a metal caused it to emit electrons.

> The light shining on a metal behaves as a stream of energy packets called photons, whose energy is proportional to the frequency of the light. (Planck had defined the formula for calculating this). When the light strikes it, the energy from the photons is transferred to electrons in the metal. If that energy is greater than what is required to overcome the forces which keep the electron in the metal, it will be released. The result is that light with a high enough frequency can knock electrons out of a metal surface.
>
> The displaced electrons are freed to move about, forming a "conduction band", and a hole is left behind where the freed electrons used to be. They are "harnessed" by the use of semiconductors with different electrical characteristics so that an electric field is generated. This field causes positive and negative charges to move in opposite directions, thus creating electric current.

The only practical use for the technology was for light meters, or so it was thought, and these became commonly employed by photographers from the 1930s onwards. By 1954, scientists working at Bell Laboratories had developed a crystalline silicon cell that performed with an efficiency of 4%. Within six years the conversion efficiency had increased to 14%. The space race gave a further impetus to research; the first US spacecraft used small photovoltaic cells to power radios.

Now, the best cells operate at 21% efficiency, cells of this quality being only very recently developed, although Boeing have managed to get a 32% rating from one of their experimental cells. This was managed with multi-junction cells that sandwich different types of material to maximise the solar energy that can be absorbed. The cells also had anti-reflection coatings to minimise the amount of light bounced off them. They also used light concentrated by a factor of 100. Concentrators are unable to use diffuse sunlight and therefore are only properly effective in dry hot areas, like deserts.

Over half the photovoltaic cells in use today work best in small products where there are low power requirements. The humble solar powered calculator uses a photovoltaic cell, as do watches, small battery chargers and Japanese soft drink vending machines and many parking meters in major cities in the United Kingdom. In rural areas of Europe, photovoltaic cell-powered pumps and drinking troughs are widely used. In Japan PV-powered insect killers are used instead of insecticides on many farms.

Eventually photovoltaic energy will become more and more important. There are plans in the United States to build a PV generating plant producing enough electricity for a city of 100,000 inhabitants. The plant, naturally enough, will be located in Nevada where it is expected to generate electricity at a cost comparable to oil produced power.

In Manchester, Europe's largest vertical photovoltaic project was partly installed at the Co-operative Insurance Society Tower in 2005. When the panel installation is complete, the tower will be covered on three sides with PV cells that are expected to create 180,000 units of renewable electricity each year.

Projects like this are inevitably very heavily subsided. Governments tend to provide very high subsidies for PV for several reasons. First, the environmental cost of electricity generated is very high and PV does generate electricity in a benign way. Secondly, they do hope to "prime the pump" and get markets created in PV technology. Thirdly, they hope to attract investment in building PV cell manufacturing plants to create jobs (although when they do they frequently have to subsidise the establishment of a factory). Fourthly, they tend to listen to the multi national companies that have invested in PV technology far more than they would to much smaller companies operating in other parts of the renewables industry. Finally, I think that people find generating electricity more interesting and possibly more "sexy" than "merely" providing hot water or heat.

The Co-Operative Insurance Society Tower project received over £1 million of public funds and its overall cost appears to be around £5.5 million. It is unlikely that the cost will ever "pay back" within the lifetime of the PV cells, which are usually estimated to be around 20 years. In my mind this raises serious questions; why are such large subsides being given to commercial organisations? It is not a question that can be answered.

The University of Oregon have calculated that the true cost to a

A photovoltaic system installed in government buildings in Seoul, South Korea.

consumer of electricity generated by PV is between 50 cents and $1.00 per kilowatt hour, still significantly more expensive than fossil fuel generation.

> The problem with efficient photovoltaics is that they use broad spectrum light whereas they would operate more efficiently at only at specific narrow part of the light spectrum. Anything outside this narrow part of the spectrum is wasted and cannot be converted to electricity. Also their efficiency drops as they become older, and in very hot weather. At freezing point silicon has a maximum theoretical efficiency of 24%; at room temperature this drops to 12%. The laws of physics mean that photovoltaic cells decrease in efficiency as the temperature of the cell increases. Because of this, for a material like silicon, the operating efficiency of a photovoltaic array will probably never be higher than 20% and will most likely be between 5 and 15%. It is probably not possible for any PV cell to achieve as much as 40% efficiency, and that efficiency approaching that kind of figure would only occur by designing cells that operate at different ranges of the spectrum simultaneously.

Photovoltaic cell technology is still immature for large-scale use. Its existing uses do save on some pollution by, for example, eliminating the need for batteries in many calculators. Many places have now installed photovoltaic cells to operate parking meters and some street lighting.

Photovoltaics really become effective in "off-grid" situations. A school in a remote off-grid part of a country can use PV to power a television, thus enabling children to watch schools' broadcasts. In places where the cost of bringing power lines or building generating plants is expensive, the PV may offer a good solution which is both environmentally friendly and cost effective.

Large-scale use is still, however, some way off on the horizon but the market is beginning to develop. Global PV cell production reached 1,000 MW in 2004 and is expected to reach 1,400 MW in 2005, increasing at around 30% a year thereafter. Japan accounts for 49% of the market, Europe for around 26% and the USA around 14%. Virtually half of all PV produced is used in Japan. In these circumstances it will be of no surprise to learn that Japanese companies, like Sharp, Mitsubishi and Sanyo are planning to double their production capacity.

Despite (or possibly because of) these projected production increase, prices for solar grade polysilicon are rising and there is an active spot market in this product. The increasing demand has led to shortages of high quality polysilicon.

The most likely breakthrough that may well give us PV that is more robust efficient and cost effective than traditional PV lies in using concentrator technology. Photovoltaic concentrator units are very different from the flat photovoltaic modules such as those that cover the Co-operative Insurance Society Tower in Manchester. PV concentrators come in larger module sizes, typically 20 kilowatts to 35 kilowatts each and they track the sun during the day. The cost and machinery involved makes them more suitable for large utility installations. The concentrator industry forecasts that using concentrator tracking technology, they may reach 40% efficiency at some stage in the future.

For the foreseeable future, photovoltaics are really only cost-effective and probably carbon effective for installations where connection to the grid would be expensive or environmentally damaging. The carbon and environmental savings of using photovoltaics in small devices where they replace batteries, is very important and should be encouraged.

A school's PV panel is being cleaned in Potosi Region, Bolivia. It powers a television set enabling the pupils to receive educational broadcasts.

Digina in the Niger has a small PV power plant at a Tuareg village. This brings electricity to this remote region.

5.2 Wind Turbines

Humankind has always valued the power of the wind. It has transported sailing vessels from the earliest times. The wind was turning mills to grind corn and wheat in China and Persia 2000 years ago, and Holland virtually owes its existence to the windmills that drained its marshes.

In modern times the real drive to use wind power for producing electricity came from Poul la Cour, a Danish meteorologist, inventor, and high school headmaster, who taught natural science in Denmark at the turn of the last century. La Cour transformed windmills to DC electricity generators in the 1900s. He also patented a mechanical device to stabilise the torque of wind turbines. His inventive mind should be credited with the wind turbine technology, upon which much hope depends for the future. La Cour taught wind energy to Danish "wind electricians". He inspired his students to become wind scientists; some of his best students built wind turbines for FL Smidth, a Danish engineering company, during the Second World War.

In the early 1950s Johannes Juul, chief engineer for a Danish power company, took up his old interest in wind energy which he had acquired during one of la Cour's courses in 1903. Juul was about to retire and wanted to have something to keep him occupied. He built some experimental machines and was the first to connect a wind turbine with an (asynchronous) AC generator to the electrical grid. Around 1956 Juul built the Gedser wind turbine. This became the basic design for all modern wind turbines and was for many years the largest in the world. The Gedser turbine had aerodynamic tip brakes on the rotor blades that were applied automatically by the centrifugal force in case the turbine turned too quickly. The Gedser wind turbine itself was built and financed by a power company.

The Danish wind turbine industry is now the world's most highly developed and successful. It started from amateurish beginnings in the 1970s when the oil crisis forced people to consider all alternative forms of energy, but eventually developed into a sophisticated and professional business.

Most wind energy projects in the 1970s began as private schemes, largely pioneered by individuals who based their designs on scaled-down versions of the Gedser machine. They produced wind turbines generating no more than 10 to 15 kilowatts of output. One of these pioneers, Riisager, built 30 wind turbines in series.

The author standing under a wind turbine in Aero, Denmark; the turbines are usually built less than 100 metres high to avoid having to light them under aviation rules. These turbines serve local communities.

Meanwhile, a number of innovative designs for smaller machines appeared, and politicians began to take an interest in the new developments, partly due to the energy supply crisis and partly in response to popular opposition to nuclear power in Denmark.

The Danish Parliament legislated that wind-generated electricity should be sold at a fixed percentage of the retail price of conventional electricity and provided capital grants for installing wind turbines in the 1970s. Unfortunately, this scheme was finally abolished in 1989.

In order to regulate Denmark's quality and safety of turbines, Denmark's Risøø National Laboratory, (originally established for nuclear power research) employs scientists and engineers in aerodynamics, meteorology, structural dynamics and other related research. The important work done there ensures the reliability of modern wind turbines. This is rightly a matter of public concern in Denmark where there is a healthy and vigorous lobby for all "green" issues.

In the early 1980s, the Danish power companies built two experimental machines, one pitch regulated and one stall regulated, of 630 kW each. For their time, these machines were very large compared with the conventional 10 to 25 kW commercial designs. Much larger designs were being researched in Sweden, Germany and the United States. In the 1980s, the State of California began a programme of support for wind energy development. Danish manufacturers, being by far the most experienced in this field, became the leading suppliers. The Californian market expanded dramatically, and volume production of wind turbines began.

> All wind turbines work in the same way; energy from the wind turns the rotor blades. The blades are connected to a shaft which in turn is connected to a generator. The wind's energy is thus used to turn the generator, creating electricity.

Wind turbines operate in fluctuating, relatively slowly moving air, using aerofoils to regulate the power output from the rotor blades. They are built in such a way as to reduce noise as much as possible. Modern wind turbines are three-bladed; an electrical motor keeps the rotor position upwind (that is to say on the windy side of the rotor pole).

Wind turbines have grown dramatically in size and performance

A wind farm at Éoliennes à Cap-Chat en Gaspésie. Québec, Canada.

during the past 15 years. The early machines of 25 kW with a 10.6-metre rotor diameter may still be found in Denmark. Today the most widely sold turbines have a rated power output of 750-1000 kW, and a rotor diameter of 48-54 metres. The largest machines commercially available are 2,000 kW machines with a 72-metre rotor diameter placed on 70-80 metre towers. Each 2,000 kW machine produces more energy than 135 old, 1980 vintage machines. Thus productivity has increased rapidly. Wind energy co-operatives and individual farmers own more than 80% of the 5,700 wind turbines in Denmark. More than 100,000 Danish families own wind turbines or shares in wind co-operatives.

Today, wind energy can compete with coal and nuclear energy on average kWh costs. The cost of wind electricity depends heavily on the average wind speed at the turbine site, because the energy content of the wind varies with its speed. Wind turbines should last for 20 years and maintenance and running costs are low. Wind energy today provides 12% of Danish electricity consumption. A target of 16% was met in 2003. In the year 2030 wind generated electricity will, it is hoped, provide 40 to 50% of Denmark's needs.

We cannot help but admire the Danes. They have invented and exploited a carbon dioxide free method of generating electricity.

A wind farm near near Brovst in Northern Jutland.

Export of the wind turbines and their technology has provided employment as well as environmental benefits. Although people complain about the sight of huge wind farms, it seems to us that the alternative, large fossil fuel power stations, is less attractive and more damaging.

In the near future wind farms will be located on the sea. Land is expensive to use, especially in northwest Europe and there are significantly higher wind speeds at sea. The wind at sea is often more stable with less turbulence and less wind shear. This makes it possible to use cheaper turbines, which should last longer. It is believed that sea-based wind farms will have no adverse effect on bird or aquatic life.

The UK has seen a large growth in wind turbines over the past few years. Many people have found them unsightly and simply do not want them to spoil the views of the countryside. This has led to people trying to question the value of wind farms, even in the windy parts of the United Kingdom. We have to consider the overall cost of them in terms of any damage they do including any environmental damage that may be caused, and balance that against the energy that they produce and the carbon savings and other benefits they generate.

One issue that has been raised is the intermittency of wind. There can never be guaranteed a steady supply of wind. That means that electricity will be generated only when the wind blows as storage of electricity on the scale required is very difficult, expensive and environmentally damaging.

When the wind is not blowing, generating plants have to start up to keep up supply, and starting up a plant (as opposed to one working continuously) means a large output of carbon on start up. Some commentators wondered whether the intermittency which meant starting and stopping fossil fuel generating stations, created more carbon than by simply letting them burn at a steady stream.

It is an important point. If carbon emission is actually increased then it means that fossil fuel use has been increased. If that happened there was no point to building a wind farm unless it was not linked to the grid and supplied users who were prepared to put up with intermittent supplies.

In the United Kingdom the Carbon Trust commissioned a study to look into all aspects of renewable grid electricity generation. Most researchers and professionals in this field took the view that intermittency does not produce more carbon than it saves and in fact the overall effect of balancing a supply between renewables and traditional electricity generation does reduce the carbon savings but only by a few percent.

> As with any source of renewables, the carbon savings
> depend upon which type of fossil fuel is displaced. The
> savings are highest when coal and oil are displaced, and if
> wind farms mean that coal burning power stations are
> closed, the environmental savings will be very positive. The
> British Wind Energy Association claims turbines save
> 860grams per kilowatt hour of electricity. That claim may
> be true for some fossil fuel alternatives but it will be more
> accurate to look at savings on a fuel by fuel comparison.

Of course over the past years, as we have seen the overall spare electricity generating capacity, (called by some "plant margin") has been reduced significantly. This may be a cause for concern but the reduction has nothing to do with using renewables but everything to do with post privatisation cost savings. Wind turbine operators claim that modern turbines have a lifetime of 20 years. Work done in Denmark claims that the overall energy balance of a turbine (that is to say the total energy used in its manufacture and ultimate scrapping when compared to the energy it generates means that it recovers the energy used to manufacture installation and de-installation after around four months. That figure is probably optimistic because it assumes that the turbines are built close to the places where the electricity is used and does not properly measure the energy losses in transmission.

Certainly, modern wind turbines have a very good energy balance, but perhaps nowhere near as good as their manufacturers claim. Time will tell.

Another important factor in considering the viability of wind farms is to consider how much carbon they save in relation to their cost. In February 2005, in Germany (probably Europe's most environmentally conscious country), a report concluded that they were an expensive way to save carbon, probably the most expensive way of all renewables. It concluded that it costs between £28 and £53 per tonne of carbon saved. In the UK the National Audit Office thought that the figure was more than double this amount.

> Germany in 2005 has more wind farms than any other country in the world; its wind farm electric output exceeds that of Denmark, Spain and the USA combined. In the 15 years ended 2005 Germany installed more than 15,000 turbines. 7,500 have been built since 2000; another 30,000 are planned by 2011, which if installed, will provide Germany with 20% of its electricity from this source.

Wind farms are very heavily subsidised; in the UK, it is estimated that over £1billion in subsidies will be spent on wind farms in the next five years, compared with what will probably be less than 5% of that amount on all the other renewable energy sources combined.

There are around 1,000 turbines in Britain and they provide only 700MW of electricity, which is about as much as a single conventional power station. In the next five years the Government plans to increase this figure twelve fold.

> The longer the distance that you have to transmit electricity, the more voltage instability you have to cater for. In addition, all electricity loses power in the transmission. Wind turbines produce a more unstable supply. It therefore makes sense to locate the turbines as closely as possible to the point where electricity is used.

A wind farm at San Gorgonio Pass, near Palm Springs, California.

The managing of wind energy involves very different issues than managing traditional thermal power stations. Wind turbines do not offer a stable supply of electricity, unlike thermal stations. They do not contribute to stabilising the grid frequency and grid voltage. The most serious issue (in the experience of Eon Netz, the largest distributor of wind generated electricity in Germany) is that when there have been brief, minor voltage dips, wind power plants disconnect themselves from the grid, whereas traditional power plants do not.

Faults like this in the extra high voltage grid can mean that all wind plants in a single region may fail simultaneously, putting grid stability at risk and affecting a whole region's supply.

Recently small wind turbines are being marketed for domestic users. Typically, these are roof mounted turbines, with blades measuring between one and a half and two metres in diameter that can generate electricity direct into a household supply. When the wind blows at more than three or four metres per second, they start producing electricity.

One of these small units will aim to produce 1 kilowatt an hour, when the wind is blowing and (provided the wind is blowing) will provide maybe as much as one third of the household's electricity use at that time.

There are a number of social and environmental issues that have, in the past, been a problem for turbine owners. Turbines are becoming quieter with each new design and it is claimed that they do not interfere with television signals. Some environmentalists are concerned about their impact on wildlife, particularly birds and bats, and by the infrastructure changes that they require: building roads to them; huge concrete bases; and their impact on beautiful areas. They worry that they will not produce sufficient useful electricity.

Professor David Bellamy, a well-known botanist and artist, feels that wind farms should be rightly called wind factories and he doubts if they will have any measurable benefit in reducing carbon emission and pollution. Greenpeace, on the other hand, regard them as objects of unconditional support.

I believe that wind is a useful source of energy, however it is not the ace of trumps in the production of energy. It deserves serious deployment in suitable locations. Its use has been distorted by the heavy subsidies that wind generated electricity attracts, not least of which are the annual renewable energy credits paid to their owners when the electricity generated is fed into the grid.

5.3 Heat Pumps

Heat naturally occurs in the ground and can be usefully used. The Romans used geothermal springs for bathing and cooking. In 1881, the King of Hawaii, David Kalakaua met with the American inventor of the electric light bulb, Thomas Edison. The King wanted to find out if the steam from Hawaii's volcanoes could be used to generate electricity for Honolulu. 96 years later, in April 1976, a turbo generator was installed to tap into steam at a depth of nearly 2,000 metres. Eventually, 25 megawatts of generating power was created – the first form of geothermal generated electricity.

Iceland has tapped its geothermal springs for over 70 years, using the heat for central heating and to provide hot water, but not for generating electricity. There is so much geothermal energy that Iceland can heat some of its major roads in winter from this source.

The real future for geothermal energy, however, lies with heat pumps. These work anywhere in the world and do not depend upon a suitably hot geothermal spring being located nearby, because they exploit heat differentials underground or in the air.

Heat pumps constitute cost effective and environmentally friendly

The geothermal plant in the Imperial Valley, California, USA.

technology. They move heat from the earth or from air into a building (to heat the building) or from a building into the earth (to cool the building). Although some electricity is used to operate the pumps, fans and compressors, the amounts used are not significant compared with the amounts that would be required to heat or air-condition the house by conventional means. A typical electrical heat pump will just need 100 kWh of power to turn 200 kWh of freely available heat from the environment or waste into 300 kWh of useful heat, although efficiency varies according to external weather and operating conditions.

The efficiency of heat pumps lies in the fact that they use low temperature heat created naturally from renewable energy sources. Because heat pumps consume less energy than conventional heating systems, their use reduces harmful gas emissions, such as carbon dioxide, sulphur dioxide and nitrogen oxides.

How do heat pumps work? Heat is transferred naturally from warmer to colder temperatures. Heat pumps, however, can force the heat flow in the opposite direction, using a relatively small amount of energy, usually in the form of electricity. In this way heat pumps can transfer heat from natural heat sources in the air, ground or water, or from man-made heat sources such as industrial or domestic waste, to heat a building or an industrial application. Heat pumps can also be used for cooling.

Almost all heat pumps currently in operation are either based on a vapour compression, or on an absorption cycle. The great majority of heat pumps work on the principle of the vapour compression cycle. The main components in such a heat pump system are the compressor, the expansion valve and two heat exchangers referred to as the evaporator and the condenser. The components are connected to form a closed circuit. A volatile liquid, known as the working fluid or refrigerant, circulates through the four components.

Geothermal systems typically use a loop of ultra high density pipe, which is either sunk into the ground or laid flat beneath the soil through which a refrigerant (or working fluid) circulates. The length of the loop depends on the heat requirement, the geographical location and soil conditions.

An ideal heat source for heat pumps in buildings will have a high and stable temperature, will be available in abundance and should not be corrosive or polluted. Soil and ground water are highly practical heat sources for small heat pump systems, while seas, lakes, rivers, rock and waste water are all suitable for use in large heat pump systems.

Geothermal heat pumps draw heat from a source whose temperature is moderate. The temperature of the ground a few feet beneath the Earth's surface remains relatively constant throughout the year, even though the outdoor air temperature may fluctuate greatly with the change of seasons.

> At a depth of two metres, the temperature of soil in most of the world's regions remains stable between 7°C and 21°C. In winter, the ground receives solar energy and provides a barrier to cold air. In summer, the ground heats up more slowly than the outside air.

In an evaporator, the temperature of the liquid refrigerant is kept lower than the temperature of the heat source, causing heat to flow from the heat source to the liquid, so that the refrigerant evaporates. Vapour from the evaporator is compressed by the compressor to a higher pressure which causes a rise in temperature. The hot compressed vapour then enters the condenser, where it condenses and

The geothermal power plant in Rotura, New Zealand harnessing the energy of the earth.

This is an Icelandic geothermal power station. In Iceland geothermal heat pipes are sunk under roads, keeping them free of ice in the long, harsh winters.

gives off heat that is then used to heat a hot water system, a central heating system or an industrial application. As the fluid leaves the condenser, it is expanded to the evaporator pressure and temperature in the expansion valve. The working fluid is returned to its original state and once again enters the evaporator. The compressor is usually driven by an electric motor but if the compressor is driven by gas or diesel engines, heat from the cooling water and exhaust gases can be used in addition to the condenser heat, improving the overall energy efficiency.

Industrial vapour compression heat pumps often use the process fluid itself as working fluid in an open cycle and because of this they are sometimes called vapour re-compressor heat pumps.

Absorption heat pumps are powered by heat rather than mechanical energy. Absorption heat pumps for air conditioning are often gas-fired, while industrial installations are usually driven by high-pressure steam or waste heat. Absorption systems utilise the characteristics that some liquids or salts possess to absorb the vapour of the working fluid. Commonly, water is the working fluid and lithium bromide the absorbent; sometimes ammonia is the working fluid and water the absorbent.

In absorption systems, compression of the working fluid is achieved thermally in a sealed circuit comprising an absorber, a solution pump, a generator and an expansion valve. Low-pressure vapour from the evaporator is assimilated by the absorbent. This process generates heat. The solution is then pumped to a high pressure, which forces it into the generator. There, the working fluid is boiled off with an external heat supply at a high temperature.

The working fluid in the form of vapour is condensed in the condenser while the absorbent is returned to the absorber via the expansion valve. Heat is extracted from the heat source in the evaporator where it heats space or water, or is otherwise usefully employed. A small amount of electricity is needed to operate the solution pump.

Ground source heat pumps have been used in many countries. The effect of withdrawing heat from soil and sub soil has never been thoroughly researched. The sub soil is being constantly heated and cooled. It is likely that heating or cooling over an extended period will alter the soil properties, in particular its moisture content. Ice may also be formed underground because heat is being extracted. It may also be environmentally damaging when heat is extracted from the ground.

As we have seen, heat pumps simply need a source of heat to

exploit. Many use the heat underground but air from the surrounding atmosphere, may also be used, although it is still a greatly under-exploited source.

Air-source heat pumps operate very well because air at ambient temperature is free and widely available. Air-source heat pumps, however, achieve on average a 10-30% lower seasonal performance factor than water-source heat pumps. This is mainly due to the rapid fall in capacity and performance with decreasing outdoor tempera-ture, the relatively high temperature difference in the evaporator, and the energy often spent in defrosting the evaporator and operating the fans. In mild and humid climates, frost will accumulate on the evaporator surface when the pump is operating in the temperature range 0-6° Celsius, thus reducing both the capacity and performance of the heat pump system. Reversing the heat pump cycle can defrost the coil but this is often done by other, less energy-efficient, means.

Exhaust air is a common heat source for heat pumps in residen-tial and commercial buildings. The heat pump recovers heat from the ventilation air, and provides hot water and/or space heating. Continuous operation of the ventilation or air-conditioning system is essential. Some heat pumps are also designed to utilise both exhaust air and ambient air. For large buildings, exhaust air heat pumps are often used in combination with air-to-air heat recovery units to provide space heating.

When the weather is humid, the evaporator coil, which in an air source heat pump is located on an outside wall, will begin to frost, and this reduces the heat output. An electronic controller senses frost build-up and temporarily reverses the flow to clear the frost.

Air source heat pumps are not large, involve no major ground works and are not unsightly, although they do have to be correctly located on a building and work with correctly sized radiators to achieve their optimum performance. Their fans provide some low level noise, but it is not generally thought to be obtrusive.

Ground water heat pumps can also be used because there are sta-ble temperatures of between 4° and 10° Celsius in many places. Heat pumps can tap into this heat source. In open systems, the ground water is pumped up, cooled and then re-injected into a separate well or returned to surface water. Open systems suffer from freezing, cor-rosion and fouling and are not, in my opinion, an ideal use of ground water heat. Closed systems can either be direct expansion systems, with the working fluid evaporating in underground heat exchanger pipes, or brine loop systems. Due to the extra internal temperature difference, brine heat pump systems generally have a lower performance, but are easier to maintain. A major disadvantage of ground water heat pumps is the cost of installing the heat source. If this is not properly (and expensively) done, they risk contaminating the water supply or affecting the water table.

Ground-source systems are usually used for residential and commercial applications. Heat is extracted from pipes laid horizontally or vertically in the soil. The thermal capacity of the soil varies with the moisture content and the climatic conditions. Soil temperatures fall during cold weather, when central heating is required. In cold regions, most of the energy is extracted as latent heat when the soil freezes. However, in summer the sun will raise the ground temperature.

Rock (geothermal heat) can be used in regions with negligible or no ground water, from bore holes between 100 to 200 metres deep. This type of heat pump is always connected to a brine system with welded plastic pipes extracting heat from the rock. Some rock-coupled systems in commercial buildings use the rock for heat and cold storage. Because of the relatively high cost of the drilling operation, rock is seldom economical for domestic use.

River and lake water are very good heat sources. Unfortunately they suffer from the major disadvantage of low temperatures in winter. Great care has to be taken in designing heat pumps using river and lake water, as it is critical to avoid freezing the evaporator.

In addition to heating water and buildings, heat pumps have many other related functions. Many industries need hot or warm water for industrial processes (such as whisky making) or cleansing processes (bottling plants) as well as hot water for washing, sanitation and cleaning purposes.

Heat pumps operating alone or in conjunction with other energy sources can meet these requirements with significantly less environmental damage and, provided that the government has created the right framework with appropriate incentives, good cost savings.

Heat pumps can operate in a variety of ways. They can provide space heating or water heating or both. They can operate as reversible heaters, cooling the air inside buildings. In the United Kingdom, heat pumped solar systems can meet around 30-85% of a household's annual heating demand subject to local conditions. Peak load needs to be catered for by an auxiliary system, often gas or oil but usually low cost low peak electricity. In larger buildings, the heat pump may be additionally fuelled by air expelled from ventilation systems.

In residential applications, room heat pumps can be reversible air-to-air heat pumps. The heat pump can also be integrated in a forced-air duct system or a hydronic heat distribution system with floor heating or radiators as part of a central system.

It is thought, however, that the overall cost of the design and installation of heat pumps makes them still fall short of proper cost effectiveness. With increased take-up and better product design, costs will fall and they will become more widely used.

5.4 Ocean Thermal Power

The ocean covers more than two thirds of the earth's surface. It absorbs huge amounts of energy from the sun every day. It is estimated that the same amount of energy as is contained in the whole of the world's oil reserves, (all one trillion barrels of it), is absorbed by the oceans in the tropical areas every week. If we could harness a fraction of the power latent in the oceans, it would produce enough electricity to power Europe and North America. It comes as no surprise, therefore, that people have tried to use some of this renewable energy.

Today, we can harness the energy contained in the world's oceans, as a result of three important discoverers: Fourier, a Frenchman; Clausius, who was German; and Kelvin who was Scottish. Fourier, in the early part of the 19th century, thought that heat transfer was subject to certain mathematical laws and could be predicted mathematically. He discovered differential equations from which engineers could predict how heat would move from one place to another. Clausius, in the middle of the same century, formulated theories about how heat would transfer and the relationship between heat and energy, and finally Kelvin, working just after Clausius, discovered the limitations between heat and energy although Clausius finally resolved these discoveries into what became known as the first two laws of thermodynamics.

Building on their theoretical work, in 1881, d'Arsonval, a French physicist, proposed that the differences between sea level seawater and deep-sea seawater could be used to generate power. D'Arsonval understood that the oceans of the world store the energy that they absorb from light in the form of heat. In fact, the oceans and seas tend to store hotter water near the surface and colder water at the lowest levels of the sea.

> The Gulf Stream current is a good example of
> temperature differentials in the oceans. Surface water is
> heated in the Caribbean Sea and the Gulf Stream moves
> that water on the surface to North Western Europe
> where it is critical in keeping Britain and Ireland warm in
> winter. When the Gulf Stream reaches the arctic region, it
> drops deep into the sea and returns to the Caribbean
> under itself, behaving like a giant conveyor belt. As more
> fresh water than normal is entering the Arctic Ocean as a
> result of more precipitation than normal the effect on the
> Gulf Stream conveyor will be to slow it down.

In order to harness the energy from the ocean, you can exploit the differences in temperature between the various layers of the ocean. D'Arsonval designed a way of using these heat differentials to generate electricity, but his design was never actually put into practice.

In 1929, a French engineer, George Claude, constructed a machine on the coast of Cuba that took the warm surface water and put it into an evaporator. The pressure was lowered causing the water to vaporise. The vaporised water was forced through a turbine where it produced 22 kilowatts of electricity. Cold water was piped up from lower ocean depths to cool the vaporised water so the cycle could begin again. Although Claude's machine worked, the pipe that collected cold water at lower depths kept breaking during stormy weather. Another system which Claude installed in a vessel off Brazil also failed, due to storms, but Claude had shown that the theoretical production of energy from the seas could be achieved.

We now know that temperature differentials in the oceans of more than 20°C are needed to make reasonable quantities of electricity. These differentials exist mainly at equatorial regions of the seas and in some tropical regions. The greatest sea temperature differential exists in the vast seas to the north, east and west of Indonesia.

In effect, ocean thermal energy conversion merely converts solar radiation to electric power. Converters use the ocean's natural thermal differences to produce a power-generating cycle. As long as the temperature between the warm surface water and the cold deep water differs by about 20° Celsius, a system can produce a significant amount of power. Many of the small Pacific islands have seas which are two or three degrees above this temperature differential range, as do Benin and Ghana in Africa, Cuba, Haiti as well as

Trinidad and Tobago. For these countries, ocean thermal installations are a highly promising potential source of energy.

As well as needing the temperature differential, an ocean thermal power plant also needs to be located in a place where this temperature differential exists within a range of no more than 1,000 metres. It should be in a place where it can be protected from storms, hurricanes and other weather hazards. If constructed near the shore it needs less in the way of moorings and other protective equipment.

How do ocean thermal power generators work? Most ocean thermal energy converters are designed to create electricity in warm, tropical waters. Warm water can evaporate liquids that boil at very low temperatures such as ammonia or Freon. The steam produced by the evaporation is forced through turbines to create electricity. The ammonia or Freon gas is then put in a storage tank into which cold water from the ocean floor has been forced, to turn the gas back into a liquid that in turn evaporates again, and so the cycle continues.

I should note that both ammonia and Freon have disadvantages; ammonia is a poison and if it escapes from the system it can kill people. Freon is one of the chlorofluorocarbons, (or CFC) group of chemicals, which when released into the atmosphere have severely damaged the ozone layer.

Modern designs for ocean thermal converters are still largely experimental and can only produce modest amounts of power. The largest is off Japan; it makes 100 kilowatts of electricity. Another converter off the coast of Hawaii produces 50 kilowatts of electricity.

The places that achieve the best results for ocean thermal installations are those with narrow shelves and very steep offshore slopes. Ideally the ocean floor should be smooth. This enables there to be a short cold water pipe-run, which is more efficient than long pipe-runs. It is also possible to build the power plant inland, well away from the shore. This would provide some protection against storms at sea that can damage apparatus. In many locations, turbulent waves in the surf zone can cause problems.

The main problem with building power plants on land is wave turbulence in the surf zone by the edge of the sea. Unless the power plant's water supply and discharge pipes are buried in protective trenches, they will be subject to extreme stress during storms and prolonged periods of heavy seas. The inevitable long pipe-runs are also a disadvantage in losing energy as well as involving very costly excavations on the ocean bed.

It is theoretically possible to locate plants on oil rig-style mountings on part of the continental shelf with a depth of 100 metres. The plant could be built onshore, towed to the site, and fixed to the sea bed. These plants would suffer from the disadvantages of high building and maintenance costs and high costs of getting the power generated on to shore.

It is also theoretically possible to use ocean thermal plants to provide drinking water, and harness their cold water discharge to help in the farming of edible crustaceans and fish or use the cold water discharge for air conditioning.

Ocean thermal power is very promising as an alternative energy resource for tropical island communities. Places like the Seychelles, Benin, Kenya, Cuba and the Caribbean Islands rely heavily on imported fuel. Ocean thermal plants in these places could provide cheap power, as well as desalinated water and a variety of marine crops. It seems to me, however, that the technology is still very immature and it is hard to see any real progress emerging in this field for decades. The real problem is an engineering one. The technology works well, but the installations have to be built to cope with the natural world to which they will be exposed and built at a cost effective price.

There are some factors that limit what can be done with ocean thermal power. There are very low levels of efficiency in ocean thermal units because of the small difference between the surface temperature of the sea and the temperature of the water at the seabed. Temperature differences in a range of less than 20° Celsius require an extremely large heat exchanger and this makes it impossible to gain energy efficiency, in theory, of more than 7%, while, in practice, the real efficiency level is much lower. Ocean thermal energy will only ever be viable in the warmer seas between the tropics but, provided the capital cost can be found, they are capable of producing a free and relatively environmentally friendly source of electricity.

5.5 Energy from Biomass

We have seen that the United Kingdom relies heavily on burning the fossil fuels, coal, crude oil, and natural gas, as sources of energy. We have also seen that nuclear energy is important. Fossil fuels take millions of years to form. Such reserves are finite and are being depleted every day by an increasingly energy-hungry world. Nuclear energy

A villager at Tilonia Village, Rajasthan, India crops the tree for cooking fuel.

is potentially problematical because, although the method of producing energy from nuclear fuel may be safe, provided all proper controls are taken (and no one can guarantee that these controls will be rigorously followed for thousands of years), the problem of storing nuclear waste by-products has never been satisfactorily resolved. In addition, nuclear fuel is not infinitely available; some sources think that we have only enough ore in the world to keep our nuclear power stations supplied for 70 years.

The only other naturally occurring combustible energy source that has the potential to be a substitute for fossil fuel is what is known as biomass. Biomass comprises all non-fossil organic materials that have intrinsic chemical energy content. You can think of biomass as fossil fuel's non-fossilised state. It includes all vegetation (both land and aquatic), trees, scrub, and all waste biomass such as the solid waste that the dust carts collect and the effluent or sewage that our houses release as well as animal wastes, forestry and agricultural residues, and certain types of industrial wastes. Biomass is only renewable in the sense that a relatively short period of time is needed to replace what is used as an energy resource. In this sense it is sometimes better classified as a sustainable energy source. It is certainly the source of energy that is most important to most people in the world.

Most scientific thinking rightly classifies biomass as a better source of energy that fossil fuels, because the CO_2 released when energy is generated from biomass is balanced by the carbon dioxide absorbed during the fuel's production. In other words, if you burn a tree you simply release the CO_2 that the tree had absorbed during the time that it grew and you are merely releasing the CO_2 that the tree would have yield up naturally when it decomposed after it died. This is called a carbon neutral process. For biomass to be a desirable alternative to fossil fuel, there must be two important factors present when biomass is used; it is necessary to replant (if biomass from crops is used) or otherwise regenerate the biomass in order to collect the carbon released by burning and the method of collecting and using biomass has to be environmentally sound.

In common with all energy that we use, biomass comes from the sun. Carbon dioxide is chemically converted into biomass, which in turn is used as an energy source. This process of conversion is called photosynthesis.

When carbon dioxide and water are processed with light and chlorophyll, carbohydrate (CH_2O), is created and oxygen is released. For each gram molecule of carbon, about 470 kJ (112 kcal of energy) is absorbed.

The usual efficiency level at which photosynthesis captures light and converts it into energy is only around 1%. This would be unacceptably inefficient in any man-made energy apparatus but the global energy potential of virgin biomass is very large. It is estimated that the world's renewable, above-ground biomass, that could be harvested and used as an energy resource, is much larger than the world's total annual energy consumption.

There are two types of commonly used biomass. "Woody" biomass is sourced from forests, woods and hedges. It can be the thin-

Coppice: some broadleaf trees, like hazel, willow and chestnut, grow abundant branches or "poles" when cut down to the stem. These poles can be harvested every six or eight years as a source of biomass. Here a chestnut tree is being coppiced.

nings of forests or specially grown trees like willow, which are coppiced regularly. "Non-woody" biomass comes from animal waste, (such as cattle manure) biodegradable industrial waste and food processing waste, as well as non woody high energy crops, such as maize, rape and sugar cane.

According to some research, modern biomass now represents only 3% of primary energy consumption in industrialised countries. However, much of the rural population in developing countries, that is, more than half of the world's population, is still reliant on traditional biomass, mainly in the form of wood, for fuel. Traditional biomass constitutes about a third of primary energy consumption in developing countries.

Biomass can be used in various ways and in various forms. Biomass product can be burnt direct, or after it has been compacted into pellets, used as gas, charcoal or ethanol or used in conjunction with a more traditional fossil fuel.

Many woody biomass products are formed into pellets. Pelletising involves compacting biomass at high temperatures and under high pressures. Sometimes biomass particles are compressed to produce briquettes or pellets, which have a significantly smaller volume than the original biomass and therefore have a higher volumetric energy density making them a more compact source of energy. Pellets and briquettes are easier to transport and store than natural biomass. They can be used on a large scale as direct combustion feed, or on a smaller scale in domestic stoves or wood heaters. They can also be used in charcoal production.

> It is important to understand the carbon content of biomass, when compared with its energy output. It should be understood that the pure carbon content of woody biomass is around 50%, compared with around 75% for coal and oil. However, biomass will be best used if it is used with high efficiency. Unlike the carbon free renewable technologies, where efficiency is only important in relation to cost because the energy source is free, in biomass efficiency is of paramount importance.

If biomass is not collected and left unburnt it will still release its carbon dioxide into the atmosphere, albeit over a much longer period than burning causes. The only circumstance where biomass

Timber newly cut from the forest at Simlangdalen, Sweden.

does not release its carbon is when it becomes fossilised. When the great coal seams where laid down from fossilised forests, the carbon was captured and stored, probably causing the planet to cool and life as we now know it to develop.

Simply burning it creates most biomass energy. The energy produced can be used to provide heat and/or steam for cooking, space heating and industrial processes, or for electricity generation. Small-scale applications, such as domestic cooking and space heating, tend to be very inefficient. There are very large heat transfer losses which can be as high as 90% even in such appliances as biomass stoves which claim high efficiency levels. More efficient stoves can help to reduce the heat transfer inefficiency.

On a larger scale, biomass such as wood, forestry residues and municipal solid waste, can be burnt to produce process heat or steam to feed steam turbine generators. The size of these generators is constrained by the amount of biomass available in their location. Usually this means that an electricity generator powered by biomass will have a generating capacity of less than 25MW. However, by using a biomass fuel that can be specifically grown locally, such as short rotation plantations or herbaceous energy crops, the generator's capacity can be doubled or trebled.

> Chicken litter, a mixture of straw, wood chips and poultry droppings, is another source of biomass. The Thetford Power Station in England will soon be the largest biomass power station in Europe: it will be fuelled by chicken litter.

Large biomass power generation systems can have comparable efficiencies to fossil fuel systems, but this comes at a higher cost due to the need for a specially designed burner to handle the higher moisture content of biomass. However, by using the biomass in a combined heat and electricity production system (or cogeneration system), the economics are significantly improved. Cogeneration is viable at present where there is a local demand for heat as well as electricity.

In Armagh in Northern Ireland, a wood fuelled Combined Heat and Power Unit has been operating since 1998. It has been burning wood to provide heating for a local museum and clean electricity for around 400 homes. The fuel is sourced from local woodlands and uses saw mill chips, but as there is considerable scope for coppicing in Northern Ireland, it is intended to provide farmers with grants to encourage their planting willow and coppicing the trees as a source of fuel. Wood chips are dried using waste heat from the Combined Heat and Power engine cooling system and then fed into a gasifier where they are heated. Heating takes place in a restricted flow of air, and this converts the chips into a combustible gas.

The gas is then cleaned, cooled, mixed with air and fed into the engine. 10% of the engine fuel is diesel supplied for ignition purposes. The internal combustion of the gas rotates the engine shaft, which is connected to a generator, thus producing electricity.

The engine exhausts contain a considerable amount of heat, which is recovered by diverting them through heat exchangers. The hot water is then pumped to the radiators in the museum for space heating.

The wood fuelled plant produces around 400 kW of heat and 200 kW of electricity at 415 volts. This is transformed to 11kV and carried away on the Northern Irish Electricity grid. The plant is capable of 24 hours a day continuous unmanned operation for 6 days after which the residual charcoal is removed and the wood chips replenished.

This project shows that not only can useful energy be provided using biomass in industrialised countries but also that there are very beneficial side effects. Local farmers are encouraged to plant trees and are paid for their crop. Coppicing takes place at seasons when local farmers are often short of work so the seasonal cycles of the crop fit in well with other local activities.

The thermo chemical process for converting solid biomass into liquid fuel is called pyrolysis. The biomass is heated in an oxygen-free atmosphere, or partially burnt in a low oxygen atmosphere. This produces a hydrocarbon-rich gas mixture, an oily liquid and a carbon-rich solid residue. The solid residue produced is charcoal, which has a higher energy density than its original fuel, and is smokeless. Traditional charcoal kilns are mounds of wood covered with earth. The carbonisation process is very slow and inefficient in these traditional kilns so more sophisticated kilns are replacing the traditional ones. The pyrolitic or bio-oil produced can be easily transported and refined.

> Gasification is a form of pyrolysis, carried out with more air, and at high temperatures in order to optimise the gas production. The resulting gas, known as producer gas, is a mixture of carbon monoxide, hydrogen and methane, together with carbon dioxide and nitrogen. The gas is more versatile than the original solid biomass (usually wood or charcoal). It can be burnt to produce heat and steam, or used in internal combustion engines or gas turbines, which generate electricity.

Biomass gasification is the latest generation of the biomass energy conversion processes. Research indicates that biomass gasification plants can be as economical as conventional coal-fired plants but they are dirty; methods have yet to be found to provide cleaner emissions and this remains the major challenge.

Commercial gasifiers are available in a range of size and types, and can be run on a variety of fuels, including wood, charcoal, coconut shells and rice husks. Again, power output is governed by the supply of biomass. The first gasification combined-cycle power plant in the world is a 6MW facility at Varnamo in Sweden, which is fuelled by wood residues.

There is a large alfalfa gasification combined-cycle power plant

In the Paragominas of the Amazon children extract charcoal from the charcoal heap. Charcoal is also produced by sawmills using mahogany off-cuts. Children often work 12 hour shifts for as little as $2.50 a week. Much of the charcoal is used to create heat energy to smelt iron ore. Opposite: *Harvesting sugar cane in Panama.*

in Minnesota, which is the first dedicated crop-fuel plant of its size in the world. Farmers grow alfalfa, a perennial crop. The leaves of the alfalfa plant are used for cattle feed. The alfalfa stems are dried to remove excess moisture. They then undergo the gasification process at the power station where, in the gasifier, the alfalfa stems are converted into a biofuel gas by being heated at very high temperatures. Burning the biofuel gas generates electricity. The heat given off from the burning gases runs both combustion turbines and steam turbine-generators. The power plant produces 75 Megawatts of electricity. State-of-the-art technology keeps air pollution emissions from the burning gas as low as possible.

Charcoal production is, as we have seen, a form of pyrolysis. Modern charcoal retorts (or furnaces) operating at about 600° Celsius, can produce somewhere between a quarter and a third of the dry biomass feed as charcoal. The charcoal produced is 75% to 85% carbon and is useful as a compact, controllable fuel. It can be burnt to provide heat on a large and small scale.

Ethanol can be produced from biomass materials that contain

sugars, starch or cellulose. The best-known feedstock for ethanol production is sugar cane, but wheat and other cereals, sugar beet, Jerusalem artichoke, and wood can all be used. Starch-based biomass is usually cheaper than sugar-based materials, but requires additional processing. Cellulose materials, such as wood and straw, are generally readily available but are expensive to prepare.

A process known as fermentation produces ethanol. Typically, sugar is extracted from the biomass crop by crushing it, mixing it with water and yeast, and then keeping it warm in large tanks. The yeast breaks down the sugar and converts it to ethanol. A distillation process is required to remove the water and other impurities. The concentrated ethanol is drawn off and condensed. This can be used as a supplement or substitute for petrol. Brazil has a successful, industrial-scale ethanol project, which produces ethanol from sugar cane for blending with petrol. In the USA maize is used for ethanol production and then blended with petrol.

Using biomass in conjunction with traditional fuel, or co-firing, is an established means of energy production. The biomass involved is usually wood chippings, which are added to the coal with wood as 5 to 15% of the mixture, and burnt to produce steam in a coal power station. Co-firing is currently well developed in the United States but the electricity generating companies are studying the effect of the addition of wood to the coal. They need to know how it affects specific power station performance and whether any problems will arise from its use.

Biomass is a useful form of fuel and in some parts of the world the only form available. It is an attractive source of energy because it is sustainable and renewable and has special features. However it is not a complete answer to the problem of creating energy in a pollution free, benign manner. It is essential to consider the complex interrelations between the important factors in the following areas before we urge the increased use of biomass.

Biomass has a relatively low energy density. Transporting it is expensive and consumes energy. This means that for biomass to be efficient the biomass processing and the energy conversion process must take place close to the source of the biomass.

Incomplete combustion of wood produces organic particulates, carbon monoxide and other harmful gases. If high temperature combustion is used, oxides of nitrogen will be produced. If natural forests are more widely used for biomass energy there will be more deforestation with serious ecological and social ramifications. This is currently happening in Nepal, parts of India, South America and in Sub-Saharan Africa.

There is a potential conflict between using land for biomass energy and using land for food production. Some biomass applications are still too expensive to use to generate power compared with, for example new, highly efficient natural gas-fired combined-cycle power stations.

The production and processing of biomass can involve significant energy expenditure, such as on fuel for agricultural vehicles and on fertilisers. This can make the energy created less than the overall energy expended in its production.

Having weighed all these features, what do we find? There are several environmental results of producing and consuming biomass energy. First, burning less fossil fuel is environmentally beneficial. Using fossil fuel and biomass together as a fuel to generate electricity in dual-fuel combustion or co-combustion power stations, results in fewer undesirable emissions.

Secondly turning household rubbish into fuel reduces the environmental problems of landfill sites. Biogas from landfills, and refuse-derived fuel, industrial wastes (such as "black liquor" generated by the paper industry) can all be burnt to produce heat, steam, and electric power. This must be environmentally beneficial as it dis-

The waste recycling plant in Croeberne, West Saxony, Germany. Every day waste is mechanically or biologically treated to recycle and to produce energy.

poses of waste usefully, as opposed to consigning it to a landfill site.

Whilst some environmental impacts may be beneficial, this is not always the case. If so called "virgin" biomass is grown specifically for harvesting as a dedicated energy resource, we destroy the biomass in creating energy but grow new biomass to replace it. If more biomass is harvested than is grown, a biomass system is not capable of continued operation as an energy plantation. Furthermore, the environmental impact of burning biomass to generate power is often deleterious because the amount of CO_2 removed from the atmosphere by photosynthesis of the biomass is then less than that needed to balance the amount of biomass carbon emitted from the cropped vegetation. In these cases virgin biomass is not renewable; its use as a fuel results in a net gain in atmospheric CO_2.

There is a real problem with biomass. Some scientists now consider that biomass burning contributes even more global warming than fossil fuel consumption. The reasoning is this: because terrestrial biomass is the largest known method of removing atmospheric carbon dioxide, by photosynthesis, overall loss of living biomass has a profound effect on atmospheric CO_2 build up. Population growth and increased land use due to urbanisation, converting forest to agricultural and pasture land, road building, destroying rainforests and large-scale biomass burning all contribute to the build up of atmospheric CO_2 at a rate that is much larger than fossil fuel consumption. Carbon dioxide is certainly the largest single factor responsible for global warming.

How, then, can large-scale biomass energy use be justified? Of course it cannot be justified unless we replant much more than we consume. Every crop must be replaced. We must probably go further than this and create new biomass growth areas. The forests are the largest, long-lived, global reserve of standing biomass carbon and replacing what has been denuded in South American, South East Asia and Africa would go a long way to help restore the ecological carbon balance, and to prevent or reduce atmospheric carbon dioxide build-up. Having replanted the rain forests it would, I believe, be foolish to cut them down again.

Biogas is an important part of biomass energy. When bacteria degrade in the absence of oxygen a gas is produced comprising methane and carbon dioxide. The process of the degradation of bacteria producing this gas is called anaerobic digestion. There are many processes that create biogas. It can be created from pig waste, cattle waste, chicken effluent, and even the waste processes of potato processing.

Biogas should be distinguished from the rest of biomass energy sources because using biogas does reduce one of the most pernicious greenhouse gases – methane. Most landfills and farms produce methane as an unwanted product. It is 20 times more potent than carbon dioxide as a greenhouse gas. Methane fuelled power stations, such as Holsworthy, in Devon, collect slurry from nearby farms and burn the methane gas released to generate electricity. Plants like Holsworthy have an important role to play in alleviating the effect of methane from human activity on global warming.

Biogas is a potentially critical source of energy to rural economies, particularly in developing countries. It has the advantage of creating a decentralised energy source using in the main locally produced waste product. In Pura Village Kunigal Taluk, Tumkur District, Karnataka State, South India, biogas has been used to create energy for many years and an energy producing large scale plant provides energy for the whole village.

Pura has about 500 people and around 250 cattle. Every morning villagers deliver cattle dung to a central plant. After being

Biogas is an increasingly important source of gas. Here it is being generated by cow dung in Andhra Pradesh, Chittoor District, in the Rishi Valley, India.

weighed and attributed to the person bringing it, it is mixed with water to create slurry.

Villagers who want the slurry can use it as fertilizer, otherwise the slurry is poured into a sand bed, which filters the slurry and releases biogas. The biogas produced is approximately 60:40 mixture of methane (CH_4) and carbon dioxide (CO_2) produced by anaerobically fermenting cellulose biomass materials in the cattle dung. This is fed into a diesel generator designed to operate on a mixture of biogas and diesel, and electricity is created. The villagers pay for their electricity but are paid for their cattle dung deliveries.

Traditionally, cattle dung is carefully collected in India and used as fertilizer, except in places where villagers are forced by the scarcity of fuel wood to burn dung-cakes as cooking fuel. The biogas plant at Pura yields sludge fertilizer so the biogas fuel and the electricity it generates is a valuable additional source of energy for the village.

Most biogas programs in the developing world are family-size biogas plants rather than plants serving a community. The amount of biogas that these small units produce is more suited for cooking than for running an engine and generating electricity.

In Indian village biogas systems, the gas is stored in an inverted

Crop of industrial hemp (cannabis sativa). The seeds produce oil that is also used in the manufacture of lubricants, paints, soap and detergents. Hemp has a great potential as an ingredient in biofuels.

drum that floats in a cylindrical pit. This delivers gas at uniform pressure and is very suited to gas produced by cattle dung. The drum is usually made of steel – reliable but costly. Similar installations in China store gas in a cement or similar constructed dome. This is cheaper but less reliable and prone to leaks.

It is of course possible to develop fuel from crops and this is now known as *biofuel*. The most important biofuel is ethanol, which is produced on large scales in Brazil and the United States. The world produces around 22 billion litres of ethanol a year. It is mostly used blended with petroleum fuels, and as a small percentage additive to diesel oil, known as "biodiesel". It is estimated that within 15 years the earth could easily produce over 30 billion litres of ethanol from cellulose – a raw material found in plants. When you consider that oil is produced at nearly four billion barrels a day and that a barrel of oil contains about 159 litres, you will appreciate how small bio-fuel production really is. If we were to use a tonne of ethanol instead of a tonne of fossil oil we would benefit in savings of carbon dioxide which some sources estimate at nearly two and a third tonnes.

> Minnesota is a state with a long tradition of growing
> arable crops. It has recently commenced a program to use
> its arable crops to provide energy. It has done this by
> blending its petrol (it uses over 2.3 billion gallons of it
> each year) with 10% ethanol that is produced from crops.
> There are 13 ethanol plants in the state producing over
> 325 million US gallons of ethanol each year all from
> agricultural sources

So, biomass in all its forms has its uses and advantages and for most of the earth's population is the most important source of energy. The real issue with woody biomass is the relationship between the carbon it emits and the carbon it generates. Like the curate's egg, it may be good in parts. If we can ensure that we burn less carbon than we grow, we can achieve a balance which while not contributing to climate change may be neutral to it.

Non-woody biomass, particularly from anaerobic digestion, is extremely valuable in the fight against climate change because it uses the methanol that would normally leak into the atmosphere and replaces it with a significantly less pernicious green house gas.

Biomass may be a sustainable and renewable energy source but that in itself does not mean that it is a benign energy source, and we should strive for sources which use fewer resources with less carbon generation.

5.6 Fuel Cells and Hydrogen

Hydrogen (H_2) is the smallest molecule but small is powerful because hydrogen is the most abundant element in the universe. It probably makes up around 75% of the Universe. Scheele discovered it in 1771, calling it "fire air" and independently Priestley discovered oxygen three years later in 1774. The French chemist, Lavoissier, collaborated with both men and took research further until he was guillotined. Henry Cavendish discovered and described hydrogen's qualities. Jacques Charles a few years later applied hydrogen in balloons, making the way open for Montgolfier's' first flight.

At this stage of discovery either a lawyer or a clergyman always seems to appear in the story. In the case of hydrogen, the Rev. W Cecil published a lengthy treatise called "On the Application of Hydrogen gas to produce moving power in machinery operated by pressure of the atmosphere upon a vacuum caused by explosions of hydrogen gas and air." Cecil thought that when hydrogen mixed with air is lit, a large partial vacuum is produced. The air rushing to fill the vacuum after the explosion could be harnessed as a force in the same kind of way that a steam engine works. By the time Cecil had described his proposed engine, breaking water into oxygen and hydrogen by passing an electric current through water (electrolysis) had been discovered, so Cecil's idea was not impractical.

Nothing much happened with hydrogen as a source of energy until the late twentieth century although Jules Verne got Cyrus Harding to claim in *The Mysterious Island* that "when coal was exhausted, water, decomposed into its primitive elements . . . and decomposed doubtless by electricity will then become a powerful and manageable force."

Experimentation occurred, over the years, including Zeppelins, a German hydrogen-powered train, a submarine, hydrogen-powered torpedoes and a proposal for a Rolls Royce hydrogen-powered aircraft. Work on developing a stable and substantial source of energy from hydrogen did not really start until the 1970s.

By the twenty first century hydrogen energy research and development was being carried out on all continents by many national governments and dozens of multi-national corporations. When they started their work, they found themselves drawn to the research and work of a lawyer, Judge Sir William Grove, a Welshman who invented something that seemed to have no practical application 130 years earlier.

Grove had an enquiring mind. He developed electro chemical cells, which we now know as batteries, using zinc in sulphuric acid and then platinum in nitric acid but his real breakthrough was to invent the fuel cells in 1839. He understood that when an electric current passes through water it splits water into its components of hydrogen and oxygen. He simply tried reversing the reaction and found that by combining hydrogen and oxygen he created a modest electric current and water.

His experiments produced some electrical energy but the results were inconsistent and it was difficult to create a reliable source of energy. He speculated on whether commercial energy could be produced if hydrogen replaced coal and wood as an energy source and this vision founded the development of modern fuel cells.

A fuel cell operates like a battery in that it will produce energy in the form of electricity and heat as long as fuel is supplied. Unlike a battery it does not need recharging, nor will it run down.

> The chemistry of a fuel cell is based on simple principles. An electrolyte is placed between two electrodes. The electrodes separate the gases from the electrolyte. The electrodes are porous and have semi-permeable membranes. If oxygen (or air) is passed into the cathode end of the fuel cell and hydrogen (the lightest element) is passed into the anode end of the fuel cell, in principle a hydrogen electrode and an oxygen electrode are established. The hydrogen is oxidized releasing electrons so that the anode has a surplus of electrons, which causes a negative charge. This process is speeded up by using a catalyst. The oxygen is reduced in the cathode, consuming electrons, which causes a lack of electrons and consequently a positive charge. By connecting the anode and the cathode, electrons are caused to flow from the anode to the cathode, and electric current is created. The electrons create a separate current. Water vapour is produced as a by-product.

Hydrogen is not an energy source like coal or oil, but a clean energy carrier, like electricity. It can be stored and converted into different forms of energy. It has been used as rocket fuel, in weather balloons and in oil refining. It is one of the simplest fuels available to us.

Hydrogen remains potentially the cleanest source of power, but the difficulties of extracting, storing and transporting it around (we are all aware of what happened to the hydrogen air ship *Hindenburg*) have yet to be fully solved.

Grove's prototype fuel cell worked but seemed to have no practical use. Further development work was done by Francis T Bacon, a descendant of Francis Bacon, who in 1959 created a five kilowatt fuel cell that he used to power a welding machine. However, the use of fuel cells was almost wholly experimental until the United States Space Program decided that using fuel cells in their spacecraft would be cheaper than solar power and less risky than nuclear power. As a result NASA developed sophisticated fuel cells to furnish power to the Gemini and Apollo craft. When the space shuttle sent manned crews into space for long periods, fuel cells not only provide the power

A pilot thermal hydrogen powered fuel cell station in Hamburg, Germany.

but also the by-product, water, was used by the crew for drinking.

At one time it was thought that fuel cells would only run on hydrogen but some research projects, particularly at the University of Pennsylvania, have created a cell which uses the hydrogen naturally occurring in hydrocarbons like petrol, diesel or natural gas. The classic fuel cell image – of a car with a huge hydrogen bag on its roof – has become a thing of the past.

Fuel cells provide a very clean form of power. Although pure hydrogen delivers no emissions when used in fuel cells (merely water vapour), petrol, methanol and diesel do deliver some emissions; however, these are significantly less in terms of power generated. This is because the process is chemically based, rather than utilising combustion. Waste products are created, but the amount of pollution is less than 10% of that produced by an ordinary vehicle.

I expect that fuel cells will ultimately have their greatest use in powering motor cars. They would appear to be more powerful than their main alternative – battery powered cars, and more energy-efficient, and would allow vehicles to travel further between refuelling.

Car manufacturers are already bringing into production vehicles that use fuel cell technology. Although fuel cell technology is still immature, scientists should be able to overcome all the teething problems, provided that there are incentives for consumers to use

The engine of the fuel cell powered Ford Focus.

fuel-efficient cars or legal requirements for lower car emissions and there is investment in research.

All the world's leading car makers are experimenting or developing fuel cell powered cars. There are already on the market cars like the Toyota Prius which uses a combination of electrical and diesel power and claims to save around a tonne of carbon dioxide a year when compared with normal usage of another family car.

Ford is bringing their fuel celled Focus FCV car into production. It plans to use hydrogen as a fuel in a fuel cell. Drawing on air and mixing air with hydrogen in a Ballard fuel cell, it will create electrical power to drive the car and water vapour as a waste product although the car will also use traditional petrol in a hybrid concept.

Ford is also involved in development of fuel cell driven motorcars. Their TH!NK FC5 – a family sized car – is powered by a Ballard fuel cell which uses methanol. The engine is located under the floor, creating a huge amount of passenger and luggage space.

One leading car manufacturer, DaimlerChrysler, expect to have an engine in production shortly. They have developed a concept vehicle based on the Jeep Commander with a hybrid fuel cell system. In this, the engine is powered by a fuel cell and battery; the fuel cell runs on petrol. The Mercedes S class model has a fuel cell as a compact auxiliary power unit. Additionally, work is being developed

Filling a car with hydrogen. Opposite: *On its way to Covent Garden, crossing Tower Bridge in London the RV1 hydrogen powered bus negotiates the traffic without polluting the atmosphere.*

on the NECAR, a fuel cell van. They have produced four generations of the NECAR so far and the fifth generation NECAR is to be unveiled shortly.

General Motors, in the United States, have developed the Precept concept car. Electricity from the fuel cell drives an electric motor on the front axle. The rear axle is separately powered. GM expects that the Precept will achieve the equivalent of over 100 miles per gallon and that they will go into mass production with a fuel cell vehicle.

BMW in Europe have a joint venture with DELPHI Automotive Systems the purpose of which is to develop fuel cell vehicles using solid oxide technology. BMW plans to fit hydrogen fuel cells that will power their series 7 electric saloon cars and fork-lift trucks.

Peugeot Citroën, Renault, Volkswagen and Volvo all have advanced fuel cell projects. Toyota and Nissan plan to introduce their own commercial fuel cell vehicles. Mazda too have produced a fuel cell concept car.

The car manufacturers are now engaged in a race to use fuel cells in cars. At the moment hybrid vehicles, like the Toyota Prius, are on the market and pure fuel cell vehicles cannot be marketed in mass

until there is infrastructure delivering hydrogen in places where car drivers can easily and reliably re-fuel. This will happen when the market reaches a critical mass but getting the market to that mass is going to be difficult, because we are in the realms of chicken and egg arguments.

The prospect of all road transport no longer burning petrol or derv or at least burning only 10% of the amount of fuel that they do at present, no longer polluting the air we breathe, and creating as by-product simple water, is very attractive. There may be other problems created by pumping excessive water vapour into the atmosphere (and after all water vapour is the most pernicious green-house gas) but we all instinctively feel that filling the air with water must be safer for us and for our children than the unkind gases that motor vehicles presently deliver.

At the moment fuel cells are expensive to produce in relation to the power that they provide. This, however, was also the case with computers until they went into mass production. Fuel cells in vehicles are likely to be a great source of carbon reduction in the future. Like all new technologies there are bound to be issues. The effect of putting water vapour into the atmosphere is untested. The issue of safely storing and compressing hydrogen still needs thinking through but on the whole fuel cells creating electricity by using hydrogen and oxygen and pushing out water as a waste seems to be a better alternative than drilling for oil, refining it and burning it with all the emissions that oil based fuels provide.

It is important to appreciate that all is not as rosy as it may seem with fuel cells. The Massachusetts Institute of Technology studied the prospects for green fuel cell vehicles and came up with an alarming report for those who think that fuel cells are going to be a panacea. MIT are concerned that even the most advanced hydrogen fuel cell car in fifteen years time is unlikely to be the greenest option. They point to the environmental cost of producing and distributing hydrogen and feel that these may well be far more expensive, in environmental terms, than present estimates lead us to believe, because it is important to consider the emissions caused in making and delivering the fuel as well as those in using it. MIT do think that ultimately hydrogen will deliver good and safe environmentally friendly car fuel but expect the time frame to be in the order of 50 years.

A better and safer way of reducing car emissions in the short term would be to legislate against or highly tax vehicles, like four wheeled driven ones and cars with massive engines. For household or domestic purposes using these vehicles are simply unjustified and very damaging.

5.7 Tidal and Wave Energy

Wave power is waterpower created by the wind and the wind is created by the sun. Like most forms of energy on our planet, energy from the sun is converted into wind power and the winds create the waves. The energy in waves is significant and could make an important contribution to the world's energy supply if we could discover an economical way of extracting it. The tides, caused by a mixture of gravity, the pull of the sun and the moon and the Earth's centrifugal force, contain plenty of energy worldwide even if we limit our use of the tides to places that have a tidal range of four metres or more. It is a shame, therefore, that wave and tidal power are so hard to extract.

There are important differences in the availability of energy derived from tide, compared with energy derived from waves. Tidal power has hourly to fortnightly variation patterns but does not vary with the seasons. There is limited potential to provide a steady stream of energy, for this reason.

Wave power has high seasonal variability and the amount of wave power available varies greatly from one place to another. In

We can use the waves as a source of energy.

Britain, the highest tidal speeds occur in spring tides and in neap tides. The world experiences spring and neap tides over a 14 day cycle linked to the relative positions of the moon, the earth and the sun. In Britain, we usually experience two high tides and two low tides each day, at about 12hours 25 minute intervals. This means that the peak in tidal velocity happens just four times a day.

Although work on wave power commenced in Japan in 1945, the real impetus for research, as with most of the other renewable technologies, came with the oil crisis in the 1970s. Work carried out in Norway and in the United Kingdom enabled more and more efficient designs to be conceived. In 1982 the first tentative British designs were produced. These take various forms, but all of them involve water pressure, which is converted in special turbines, into electricity.

So far the technology has produced self-powered buoys for use in navigation and weather monitoring. Power plants that use waves, built in Kaimei, Japan, and near Tsuruoka City, also in Japan, have proved disappointing in that their electrical output has been more expensive to produce, and, when produced, has actually been substantially less than was originally predicted.

In 1974, Professor Stephen Salter and his colleagues invented the Salter Duck at Edinburgh University, so called because it resembles the head and beak of a duck. The duck is positioned (with others) parallel to incoming waves, so that the beaks face them. The force of the waves causes the ducks to pivot and the pivoting drives a hydraulic pumping system that powers a generator.

In Madras, India, the Indian Institute of Technology has built a wave energy plant that generates 25 kilowatts for the four months when the sea is relatively calm and 75 kilowatts when the sea is rougher. In the monsoon season it can generate as much as 120 kilowatts. The success of this plant has encouraged the local authorities to design others that will be larger.

In 1985, Kvaerner, the Norwegian engineering company built a wave plant 18 miles north of Bergen, producing up to 500 kilowatts. It performed reliably for three years until severe storms destroyed it in December 1988. A smaller plant is functioning elsewhere in Norway.

> The highest concentration of wave power can be found in the areas of the strongest winds, between latitudes 40 degrees and 60 degrees in both the northern and southern hemispheres, on the eastern sides of the oceans. The United Kingdom is thus one of the best situated places for the extraction of wave power, but even here large-scale projects have been shelved for the foreseeable future as they are thought to be uneconomic.

A wave power station was built on the Isle of Islay, in the Hebrides. The generators were built by Wavegen designed in conjunction with Belfast University with financial backing from the European Union. The system has two major components. A wave energy collector and a sloping reinforced dome have been built into the rock on the shoreline. Seawater enters a central chamber and can leave it, according to the wave movements. As waves enter, the air in the chamber is compressed and this air is forced through the

*Water is pouring out from a tidal reservoir in the Old Tide Mill, Carew,
Pembrokeshire. This energy in the moving water can be transformed into useful
electricity.*

second component, a generator, which produces the electricity. Professor Alan Wells of Belfast University designed the special generator that is used. As the seawater leaves the chamber the air pressure
drops and as air is sucked back into the chamber the turbines of the
generator are designed to enable this energy too to be converted into
electricity.

In Australia, Energetech has developed a turbine that is also a
two-way device that they claim is significantly more efficient that
Professor Well's turbine. They propose to use it in parabolic funnels
to enable more wave energy to be captured and used.

Tidal energy is easier to convert than wave energy. In the right
coastal environments, usually at the entrances to large estuaries,
tidal resonance occurs. This produces a far greater than average tidal
range. Good examples are at the Bay of Fundy, Canada, (this has a
mean tidal range of 10.8 metres, the largest in the world), and in
the Severn Estuary (with a mean range of 8.8 metres). Tidal energy
technology is largely conventional and has been tried before in other
areas on a smaller scale.

The feasibility of extracting power from the tides in the Severn

Estuary was explored in the 1970s but shelved as being uneconomical. However, today there are proposals for a tidal power plant, which, if built, could produce an estimated 8,600 megawatts incorporating 216 turbines – this would be roughly 7% of the United Kingdom's total electricity demand.

> The only major tidal power stations in operation today are the 240 MW facility in St. Malo, France, finished in 1966 and comprising a large range of experimental facilities, and a 20 MW unit in the Bay of Fundy, Canada, dating from 1984.

Most designs for tidal power stations involve the use of large artificial man made barrages across an estuary or tidal region in order to control and harness the natural flow of the tides. Building barrages is only really economically viable in relatively shallow seas and estuaries. Sluice gates and generators are located along the barrage and, using pumps to increase the tidal water captured, can provide a significant amount of increased energy out put. There are similarities in concept between these kinds of installations and low-head hydro electric dams, because the same technology is used in each case.

If we need to build huge civil engineering structures to harness the tides we must also consider the environmental cost. Tidal energy produces carbon free power when it is installed but building these structures leaves a large carbon footprint.

Neither wave nor tidal power is cheaper to produce than conventional power sources. Wave and tides cannot produce steady rates of electricity at peak demand times, and the energy cannot be stored conveniently or economically.

The development of tidal and wave power is not proceeding as quickly as it should. In Britain, (which should be in theory one of the best countries in the world for exploiting wave and tidal power) a research group was formed during the energy crisis of 1976, but the research was shelved in 1983 when it became apparent that no large-scale plants would be economic in the short term.

While wave energy is used successfully in very small scale applications, such as powering lighthouses or navigation buoys, its short term prospects as a major contributor to large scale energy production seem to be uneconomic.

Until we accept that we cannot continue to push carbon dioxide

This tidal power plant is located in the Annapolis Basin at Bay of Funday, Nova Scotia, Canada.

into the atmosphere, and until we are forced to recognise that cost cannot only be measured in monetary terms, the likelihood of tidal or wave power playing any significant role is virtually nil.

5.8 Thermal Solar

Thermal solar is the use of light to generate energy in the form of heat. It starts with the sun. The sun is the source of all our energy. It is a sphere made up of intensely hot gases, the centre of which is unimaginably hot and has a density about 100 times that of water. The sun's activity is like that of a continuous fusion reactor, constantly emitting energy by radiation. The energy is produced by the fusion (or collision) of matter at its core; the gases of which the sun is composed contain the fusion reaction and are themselves held together by gravity.

The energy is pushed to the outer zone of the sun by convection and from there is radiated out into space at various wavelengths as light. So, what is started by a heat process is turned into a light process. Light can travel through the vacuum of space, but only lim-

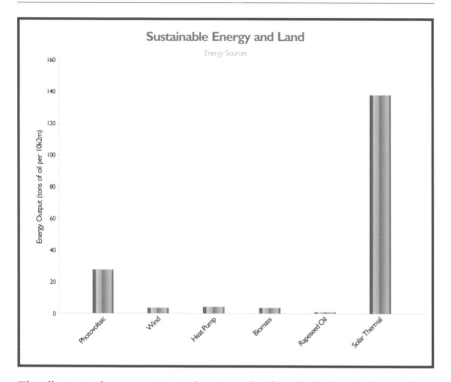

This illustrates the energy potential, expressed in barrels of oil equivalent, of a hectare of land, according to various renewable energy sources. It should be noted that it is possible to use the same plot of land for underground heat pumps, biomass culture and wind turbines, but the combined energy value is still less than thermal solar.

ited amounts of heat can. When light reaches our planet it is received in two forms. *Beam radiation* is radiation received from the sun without being scattered by the earth's atmosphere. Other energy is received in a form that is scattered by the atmosphere and is referred to as *diffuse radiation*. The scattering occurs when the radiation passes through air molecules, water and vapour droplets and dust. The two forms together are known as total solar radiation.

Above the earth's atmosphere, the amount of light being received (known as *insolation*) is constant and reliable. However, at any place on the earth's surface the amount of radiated energy received at any given point is determined by the scattering effect that I have explained and the spherical shape of the planet (which makes some parts of the world further away from the source of light than others), the stage of its rotational and seasonal cycles and the latitude of the place.

> Every year comprises about 8,760 hours (more in leap years). Half of these hours are spent in night time at any particular location and of the remaining time somewhere between 1,300 and 2,000 hours are spent in sunshine. Although sunshine is not always going to be present as a source of energy, solar radiation is present during all daylight hours and in all weather conditions.

In simple terms when radiation (or light) strikes matter in the atmosphere of our planet it causes the molecules of the matter to vibrate. That vibration causes friction and friction is heat. Thus, when you are outdoors on a cloudy day you can still get badly sunburnt, because the light is causing your skin's molecules to vibrate and the friction of vibration causes the burning.

These principles form the foundation of thermal solar but thermal solar systems developed not from any detailed understanding of the physics but initially from simple experimentation and observation.

Around 1760 Horace de Saussure, who was Swiss, noted that heat is retained and increased when a room or carriage has glass windows and experimented to see how much heat he could trap. He built a small wooden glazed box and measured the temperatures attained inside when he exposed the box to the sun. To his surprise, he found that temperatures easily exceeded the boiling point of water. Like many inventors, De Saussure was unsure what he could do with his glazed box, but felt that it would be useful one day. He was right. That glazed box was the precursor of modern thermal solar panel technology.

Up until the end of the nineteenth century, water was heated only by using stoves, usually burning wood or coal. In some cities coal gas was used, but that technology was generally unsafe. Even using coal or wood had huge disadvantages. To get warm water someone had to get the fuel carry the fuel to the stove, light it, wait for it to heat up, wait for the water to heat up and then use it. This was highly labour intensive work. It made washing harder. The body odours that existing two hundred years ago may have been quite stultifying, and of course unwashed people and clothes did harbour diseases.

EARLY THERMAL SOLAR APPARATUS

At the end of the nineteenth century the first solar collectors were developed in the United States, usually by the frontier people, who were hunting farming or prospecting. They built water tanks, painted them black and simply placed them in the sun. They were called "collectors" because they collected the light (or sun) to heat water. They were not terribly efficient. It might take ten hours to heat a small tank of water, but that was better than using cold water or wasting time chopping wood for the fire.

Clarence Kemp sold heating equipment salesman from Baltimore, became fascinated by the concept of solar water heating and used his engineering skills to design and then market, in 1891, the first commercially produced solar water heating system, building on the frontiersmen's home made solar heaters. Kemp put it on sale with some success in the north eastern states. He called it "The Climax".

Kemp's design was for a single tank, exposed to the cold night air. It worked but could be improved on and was in 1909, when William J. Bailey invented a solar water heater that exposed the collector to light but kept the water stored in an insulated tank inside the home.

Bailey's collector design was for black pipes attached to a metal sheet in a glazed box. The heating element consisted of pipes attached to a black-painted metal sheet placed in a glass-covered box. Bailey's collector heated water faster and provided hot water in the tank, which could be used the following day. He called his invention "The Day and Night" and sold over 4,000 systems in less than ten years.

This promising start was brought to an abrupt end when large amounts of natural gas were discovered in California. It was much cheaper to use gas for water heating than coal or wood, so Bailey developed a gas water heater and sold the rights for his Day and Night system to a Florida company in the 1920s.

In Florida, coal and oil were expensive and the solar systems sold well, matching the building boom in the thirties, until by 1941, around half of the homes in Florida

This array of solar collectors was installed at Bedford Town Hall to provide hot water. The panels are being mounted on A frames on the Town Hall's flat roof. (Future Heating Limited)

had thermal solar. However as energy prices declined, fewer thermal systems were sold. The flourishing solar industry in Florida was finally ended when electrical companies offered electric water heaters at bargain basement prices thereby increasing electricity usage.

However after the Second World War many European countries developed highly sophisticated solar collectors. They improved the performance of the collectors by developing special coatings instead

of using black paint, developed various ways of exchanging the heat and protecting the heat that they had collected by excellent insulation material (which stopped most of the heat escaping) and eventually vacuums (which prevent virtually all of the heat escaping).

Although vacuum collectors are fairly simple, the concept of a vacuum is relatively new, in terms of scientific history. One German scientist played a particularly important part in proving that vacuums exist and in inventing apparatus for creating vacuums. In 1602, Otto Guericke was born in Magdeburg. This was a time of great scientific exploration. Guericke studied law at Leipzig University. Like many educated young men of that time he studied mathematics, geometry and mechanics and found that he had an aptitude for what we would now call engineering.

After touring England and France he resettled in Magdeburg and planned the rebuilding of the city using his engineering skills. When he was in his mid-forties he was elected Mayor and then entered the most fruitful phase of his scientific career.

He was a great inventor. He invented an air pump, which was crucial because at that time he was looking for ways to create a vacuum. He managed to create vacuums in glass and metal vessels, provided that the material used to make the vessel was strong enough to stand the outside air pressure. He also proved that inside a vacuum, candles would not burn, animals could not live and a ringing bell could not be heard. These were unknown qualities of a previously unknown property – the vacuum.

> Guericke's experiments impressed the Elector Friedrich Wilhelm who asked Guericke to arrange a demonstration of a vacuum.
>
> Guericke built a globe made up of two small hemispheres fitted with leather washers. It was slightly larger than a basketball and could be pulled apart by a small child. He then used his air pump to evacuate the globe (creating a vacuum inside it) and harnessed eight horses to each hemisphere. After a great deal of strain the horses eventually managed to pull the small hemispheres apart. As the globe was pulled apart there was a loud frightening bang.
>
> The Emperor ennobled Guericke, and Otto became von Guericke. He died in 1686 at the age of 84. Much later, the University of Magdeburg was named in his honour.

Later, an Irishman, Robert Boyle, improved on Guericke's invention, working jointly with Robert Hooke in London. Hooke and Boyle invented an air pump they called the pneumatic engine. Boyle later became famous for his law defining the way in which a given volume of air varies inversely to the pressure on it, but alas, no university has been named in Boyle's honour.

Building on the work of von Guericke and Boyle, the most popular use of evacuated vessels was, for many years, the humble vacuum flask. For a long time the technology was only operated in laboratories for experimentation. Gradually other practical uses were found for it.

The engineering company, Dornier Systems were inspired by oil shortages of 1974 to research energy-creating applications that did not fossil fuel. The Dornier team's first invention was a flat plate solar. Trials proved that these collectors were unsuitable for what was then thought to be the main market – deserts and tropical zones. They were badly affected by corrosion and this degraded performance.

Ultimately, designers of all solar collectors got to realise that the part of the collector that absorbs energy must be selectively coated in order to absorb the light as completely as possible. We have seen that after the surface absorbs light, the light energy is transferred to heat energy. A heated absorber surface tends to emit long- wave radiation in the infrared range of the spectrum back into the surrounding atmosphere. This causes heat loss.

However, Dornier discovered that if the absorber surface is coated with a selective coating, it gains properties of high absorption and low emission of radiation. This is expressed as the ratio of absorption: emission. High grade selective coatings have absorption: emission ratios of between 10 and 15:1; they absorb a lot and emit little. The radiation that has travelled through space to a solar collector is actually absorbed by a blackbody or absorber plate. This is a black or navy blue plate inside the vacuum tube or the solar panel. (Although the absorber plate is called a blackbody, it is not black but a very deep shade of blue; it just appears black to the human eye).

The blackbody collects the radiation and, because of the efficiency of its design, emits as little of it as possible in the form of reflected radiation. But the energy has to go somewhere. The blackbody, besides being a very good absorber of solar radiation, is equally good at converting the solar radiation into heat. If the blackbody is insulated from its surroundings it will eventually cause any adjacent heat pipe to reach the same temperature. They needed to design a collector where the radiation of the sun is converted into heat, while long wave radiation back into the surrounding atmosphere is kept as low as possible. The process by which the light absorbed is converted into heat is referred to as "heat transfer".

If the heated absorber surface is surrounded by air, heat is also lost through the ordinary processes of conduction and convection. Therefore the collector must be equipped with highly efficient heat insulation to ensure there are no convection and heat conduction losses (that is to say to prevent the heat being dissipated into the atmosphere). Cold weather and cold wind can both prevent solar collectors from working efficiently.

This factory in Ziar, Slovakia, has 80 thermal solar panels providing hot water for the factory employees' showers and canteen. The picture was taken in below freezing weather when the system was providing hot water to the factory. Inset: *A ThermoSolar panel system providing water heating for a German swimming pool.* (ThermoSolar AG)

The designers of the final product had to take two problems into account. First of these was to use the properties of insolation. Traditional solar panels did not convert radiated solar energy into heat, but rather used direct heat from the sun. They only worked on sunny days in hot weather. Solar collectors, like the early American ones, that ignore radiated energy have almost no application in most parts of the world today.

Solar radiation, on the other hand, is present and measurable on cloudy and cold days and is available in sufficient quantities to be useful at all latitudes and in all climates. The second problem that the engineers needed to address was not how to collect the energy (once they had determined that there is sufficient radiation in daylight hours, even at northern latitudes, for useful applications), it was how to prevent most of the energy radiating straight back out of the panels.

Some collector designers used insulation. Today high quality insulation is available and prevents huge amounts of heat losses. Others, like Dornier, used vacuums to prevent heat loss, creating evacuated tubes of glass with a blackbody inside surrounding a heat pipe.

> On average in the United Kingdom, it is calculated that insolation is sufficient to provide 100% of a family's hot water requirements in June and July, 95% in May, over 80% in April, August and September, 60% in March and October but less than 30% from November to February. Conditions will vary from place to place, but with a normal collector installation 50% to 80% of the hot water supply should be achieved by purely solar means. Even in the short cold days of winter the collectors still work, but not as efficiently as in summer. This is because in the United Kingdom there is much less daylight in winter than in summer, as compared with places closer to the equator, and also because the angle of the sun is much lower in the sky in the winter and that makes the light received weaker in intensity. Although the solar collectors will not be contributing the 800W per square metre of output that they achieve on a bright summer's day, they will nevertheless contribute significant amounts of free hot water.

Much debate has centred upon whether flat plate panels or vacuum tubes are better for heating our domestic water. Both systems have their supporters and both systems have been around for a number of years, so it is possible to utilize actual case studies and experience in making a comparison, rather than theoretically based assumptions.

Both flat plate panels and tubes work in the same way and come in various designs with comparable efficiency. The ability to provide useful hot water is in both cases limited by the laws of physics and the demand requirement of domestic hot water.

In the case of vacuum tubes, the blackbody is enclosed within a glass tube, which is then evacuated. Light can travel through a vacuum but heat cannot, so in effect the vacuum is used to insulate the system and prevent virtually all heat loss. The glass tube, when evacuated, is connected to a metal condenser and the heat is directed to the condenser, which fits into a manifold. The vacuum seal is located where the glass tubing meets the manifold. The heat exchange fluid passes through the manifold.

In the case of flat plate panels the heat exchange fluid serpentines in pipes connected to the absorber plate. Genersys panels are constructed so that the absorber plate wraps around the heat pipe, enabling almost complete contact between the two. Other panels have the pipe welded or soldered to the absorber plate.

It may be thought by a layman that solar collectors are designed to collect as much light as possible and convert it to heat. This is actually not the case. There is no point in overheating the system because the hot water usage and storage capacity is finite and the way in which people use hot water and the times at which they use also have to be taken into account. Overheating simply reduces collector life.

Genersys, for example, uses specially designed and manufactured selective coating for its collectors, which is designed not to get too hot. There are other coatings that can be used, which get hotter but it is not helpful to make making collectors that create more heat than can be usefully used because the higher the temperature at which a thermal system stagnates and the more frequently the system reaches this stagnation temperature, the faster the ageing process of the whole system. Very hot panels can be used in systems where the demand is always going to be more than the system's performance.

I stress that both vacuum tubes and flat plate panels are viable and suitable systems for heating domestic water. If they are properly manufactured they should both give many years of useful production. In larger installations, a competent engineer should make an evaluation that takes into account all the factors that apply in the particular circumstances of the user and the location.

The very design of vacuum tubes inevitably creates certain advantages and disadvantages.

Tubes are quite easy to fit as the manifold can be mounted on the roof and the tubes carried up to the roof by one person. Flat plat panels can be heavy. Tubes perform slightly better in relation to their size; generally a

An array of evacuated tubes providing heat for water to a government building in Seoul, South Korea.

Part of a 132 panel system providing pre heated water to boilers for 452 flats in Zilina in the northern part of Slovakia. For six weeks in summer these panels provide all the hot water required. For the rest of the year the panels save fossil fuel by pre-heating water from about 7° C to about 30°C. Frost has formed on parts of the panels which are shaded by other panels.

vacuum installation needs around 10-12% less roof space than an equivalent flat plate system.

If an individual tube fails it can be replaced; there is a relatively much higher failure rate of individual tubes (compared to the failure rate of well engineered individual panels). A tube failure can be diagnosed either when fogging is apparent or frost is not visible on a tube when it is visible on other tubes in the same manifold.

Tubes are prone to over heating because it is difficult to design tube systems in a way that avoids overheating. Some tube systems try to overcome heating issues by incorporating automatic valves in the manifold. Properly designed flat plate panels, like Genersys panels, never suffer from overheating that can damage the panels or the system components because the panel is designed specially to avoid this. Overheating reduces the life of the systems.

Well engineered panels are much more robust than tubes.

The stresses caused by the expansion and contraction of the glass in tube systems (the coefficient of expansion of glass and metal are not identical) can lead to stress where the glass is joined to the condenser and sometimes stress fractures are caused, which means that the vacuum fails.

The vacuum seal, located as it is in tubes where the glass tubing meets the manifold is actually located upon the hottest part of the collector causing stress upon the seal.

The system of holding the tubes in a manifold and securing the tubes to a roof means that in windy conditions minor tube movement can create glass fractures, which lead to vacuum failure. Panels do not suffer from this inconvenience.

In snowy conditions snow tends to remain in the gaps between the tubes, reducing efficiency, whereas it tends to slide off panels much sooner that it clears from tubes.

Panels can be roof integrated and are actually generally cheaper to install in new build situations. It is not possible to integrate tubes into the roof.

Installations with vacuum tubes usually require more service calls than installations with panels owing to the more fragile construction of tubes.

Racks of evacuated tubes.

Some people like an array of futuristic looking tubes on the roof whereas others prefer the flexibility that panels bring; panels can be fitted on to a roof, in the same way that tubes can, but unlike tubes panels can be roof integrated, which I think makes them look more pleasing and fits in better with architectural designs. A large array of panels can be made to look interesting by contrasting the roof tile colouring with them or could be made to blend by using blacks and dark blues as a roof tile colouring.

In Germany, where both types of solar systems have been used for over thirty years, (in numbers that are far greater than in this country) the market originally favoured tubes and panels were considered as a second best choice. Today vacuum tubes comprise less than 10% of the market, with flat plate panels taking virtually the whole market. End users found panels just as efficient, more aesthetically pleasing but also longer lasting with fewer faults and service calls required.

Most Northern European systems are sealed. This means that heat is pushed in through a sealed pipe into a special coil in a hot water cylinder by a pump. When the cylinder reaches the pre set temperature the pump stops. When cold water is drawn into the cylinder as hot water is used, the pump starts. This way the potable water never comes into contact with the panels. This means legionella risks and panel contamination risks (scaling and corrosion due to impurities in potable water) are avoided.

Some systems do not use this indirect method of heat exchange, which means that panels need to be cleaned and serviced. Other systems are designed so that fluid in the panels drains back when the weather conditions outside are too cold, which could lead to freeze damage, or too hot, which could lead to water boiling. The water or freeze resistant fluid inside the solar panel will drain back into a small drain back bottle. The system is thus protected against damage due to boiling and freezing in a simple way.

The experience of Europeans in Germany, Austria and in many other countries has led most thermal engineers to believe that drainbacks are now old fashioned technology. In Germany, most engineers take the view that sealed pressurised systems using indirect heat exchange are better, more robust and more efficient. There are no installation problems with sealed pressurised systems although there is a requirement for high quality methodical workmanship during installation.

Freezing is an important issue: drain back systems are designed

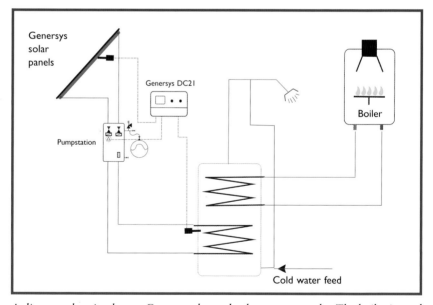

A diagram showing how a Genersys thermal solar system works. The boiler is used as a back up for times when there is not enough light energy. A typical system in the UK saves around three quarters of a tonne of carbon dioxide each year.

This thermo siphon system in Kato Amiandes, Limasol District, Cyprus, has the cylinder mounted above the panels, in order to allow gravity to feed the heat into the cylinder. This is a cost effective system in places where potential wind damage is very unlikely to happen.

to drain all the water inside the panel away to avoid freezing. Non-drain back systems use edible glycol to avoid freezing, with a gel point of −30°C. It is food safe and benign, not like traditional antifreeze. Drain back systems cannot work below freezing point, whereas pressurised systems are designed to work at below freezing conditions, such as those experienced every winter in central Europe.

Over-heating has been traditionally associated with thermal solar installations and so drain back systems do not work when the weather is hottest. Non-drain back systems work whatever the outside temperature, providing there is enough light.

Thermal solar has been used effectively for many years. As we have seen, the deployment is directly related to the cost of fossil fuel energy; the more it rises the more likely thermal solar will be used. It is a mature technology which has been used for nearly a hundred years. People like it, and feedback shows a high level of customer satisfaction.

Solar Panels being installed into a house in the course of construction in the Highlands of Scotland. Photo courtesy of Everwarm Services Limited.

THE PAY-BACK ISSUE

Many people ask "What is the pay back?" for solar thermal systems. The same question is usually asked of every renewable energy system. Well, first and foremost it does have a payback. If you buy coal, oil, electricity or gas the supplier does not offer you any pay back at all. It's all an expense.

If you take a proper view of payback, factoring in future energy increases, the savings in boiler servicing and the ability to make a boiler last longer, most solar thermal systems offer a pay back of between 6 and 15 years, depending on the fossil fuel displaced. If the product is built as part of the housing infrastructure it will last for around 30 years with minimal maintenance. If the product is built into a home in the course of construction the pay back period declines rapidly.

Solar thermal systems still need fossil fuel back up, but this does not affect the pay back. When fossil fuel systems are installed no-one looks for a pay back. When a thermal solar system is installed any pay back, even a long one, is an excellent reason for installing it.

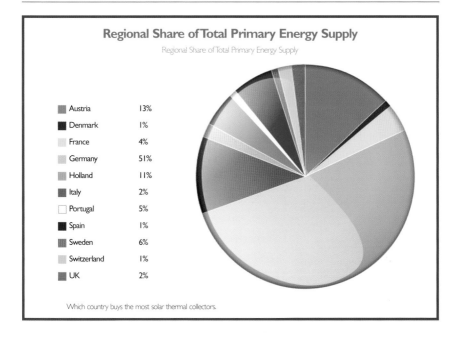

Regional Share of Total Primary Energy Supply

Regional Share of Total Primary Energy Supply

Austria	13%	
Denmark	1%	
France	4%	
Germany	51%	
Holland	11%	
Italy	2%	
Portugal	5%	
Spain	1%	
Sweden	6%	
Switzerland	1%	
UK	2%	

Which country buys the most solar thermal collectors.

This graph illustrates which countries bought the most solar thermal panels in Europe in 2005, showing the German predominance in this field of renewable technology.

Whichever type of thermal system is chosen, the most important thing is that a solar thermal water heating system is installed. Cost savings and "pay back" issues may be important to some buyers but equally important is that a single system will save between a third and three quarters of a tonne of carbon dioxide each year, and thus will benefit the planet as well as the end user.

Finally I should mention solar furnaces and concentrators. Furnaces use mirrors and lenses to concentrate direct sunlight into a small space and produce very high temperatures. At Odellio, in France, a solar furnace is used for scientific experiments. 11,000 flat mirrors on a hill are positioned to track the sun and reflect its rays onto a large curved mirror which covers one side of a ten-storied building. The curved mirror focuses the sun's rays onto an area of less than one square metre in a tower where temperatures achieved can reach 33,000 degrees Celsius.

Solar concentrator systems use lenses or reflectors to focus sunlight onto the solar cells or modules where the intensity is usually

Solar concentrators used to generate steam for cooking at Rajasthan, Mount Abu Brahma Kumari Ashram India.

This installation needed a south facing orientation so the installers built one. (Celtic Solar)

increased to between 10 to 500 times. Concentrator system lenses are unable to focus diffuse or scattered light, limiting their use to areas, like desert areas, with a substantial number of cloudless days on an annual basis. They can only use direct sunlight and therefore must track the sun, and they also must have some cooling system.

Thermal solar for water heating is mature, robust and relatively simple and cheap. It really ought to be one of the primary measures that are used in every climate. Compelling its use (as some countries like Cyprus do) makes sense if people are to begin the process of genuinely reducing emissions, particularly from the home.

Chapter 6

Combining technologies

As we have seen, most of the "green" or "renewable" sources of energy have one major disadvantage over fossil fuel. Fossil fuel, despite its finite nature, can be stockpiled and used as and when necessary. Every country has depositories where oil, gas or coal is stored and then distributed or fed into a power station for conversion into electricity. In developed countries, like the United Kingdom, the net work is sophisticated and highly integrated. Most people buy gas delivered into the home or, if they are off the gas grid, they can arrange for a virtually seamless supply of heating oil. Most homes are on the electricity grid and so the electricity generating companies have to cope with surges in demand and the inefficiencies are hidden from the consumers; they simply turn the switch and the power flows.

There is always some element of intermittency with renewable sources of energy. The sun does not shine at night; the wind does not blow all the time. In fact, the oil and gas will not be available for conversion into fuel for as long as the sun will shine and the wind will blow and the tides will turn. But dwindling supplies of fossil fuel may be some way ahead in the future and we all have a tendency, like Mr Micawber, to believe that something will turn up and perhaps this optimism in the future is preventing us from properly exploiting the renewable technology that presently exists.

There is great scope for combining renewable sources of energy with fossil fuel sources. We are already doing this to a very limited extent, because the electricity generating companies are using wind turbines as well as oil and gas and coal burning generating plants, but as we have seen, the actual energy output is limited and intermittent.

I think that most useful deployment of renewables probably lies in small scale applications where the renewable energy source is combined with other sources. However, I shall first explain some of the larger scale options of combining the various technologies that are available.

District heating systems are a good example of combining different energy sources. In the United Kingdom, most people think of district heating as a small scale operation, possibly powering a few homes and a few public buildings. In Scandinavia, district heating is much more widely used and the projects are much larger.

At their best, district heating systems are designed to provide the sole source of heat energy in large numbers of homes and buildings entirely from sustainable sources in the most environmentally and cost effective manner possible. They can heat thousands of homes, schools and other public buildings and provide clean green energy for people in their ordinary every day lives, everywhere drawing on local sustainable or carbon free energy sources.

> The world's largest district heating systems is on the Danish island of Ærø and serves fourteen hundred homes in Marstal. The same island has a more modern system which serves 115 homes in the village of Rise. These communities are now empowered in their environmental decisions and have energy independence and security. They operate on the principle that a single district heating system should serve a single identifiable community and should be managed by that community.

A district heating system has to be designed to generate heat. On Ærø, the system does this in two ways. The first way is to create heat energy from thermal solar collectors arranged in a solar field. This is on a very much larger scale than a traditional thermal solar system but exactly the same principles apply. Light energy is converted into heat energy in solar collectors arranged in a large field. The energy is used to heat water, which is stored in a huge tank or in an underground reservoir. As we have seen, light energy is not always present so for district heating systems to work all year round they need to have a back-up source of energy, just as a home with thermal solar needs a back-up source.

In the case of Ærø, the system is backed up by heat generating from a biomass boiler using locally available biomass fuel. If, for example, wood pellets are available locally, then a wood pellet bio-

A sea of district heating solar collectors on the island of Ærø in Denmark. These are the world's largest district heating systems. Note the variety of collectors used.

mass boiler is used. If straw is available, a straw biomass boiler is used. The boiler provides the heat at times when the solar field has insufficient light to provide heat up to the necessary standards.

The system uses a variable electrically operated pump to pump the water around a thermal ring main. The thermal ring main runs underground and is made of super-insulated steel tubing. The heat is pumped into homes, directly into radiators and into the heat exchange coil of hot water cylinders.

There are many advantages of district heating systems. They are self- contained and provide energy independence. Local biomass can be grown to supplement solar energy. They provide carbon neutral (more or less in the case of biomass) and carbon free (in the case of solar) energy for households.

The environmental savings of district heating are very impressive; a yearly solar production of 3,600 mWh saves 400,000 kilograms of oil, over a thousand tonnes of carbon dioxide, 1,400 kilos of nitrous oxide and 1200 kilos of sulphur dioxide. An equivalent saving would require solar water heating-only installations to be

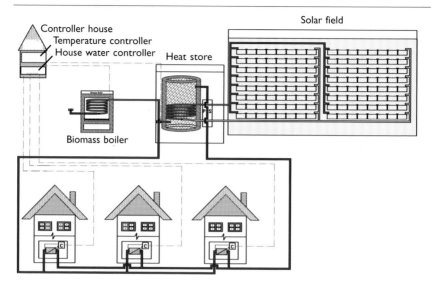

A diagram of a typical solar power district heating system.

installed in about 1500 homes. If the United Kingdom installed only one hundred 3,600mWh District Heating Systems, the overall environmental savings would be 40,000,000 kilos of oil a year.

It is not difficult to find a hundred rural communities that can be served in this way. The carbon savings would be much higher than for wind turbine. Communities with district heating will develop an increased environmental awareness and this will probably lead to more positive environmental behaviour in those communities.

I believe that many communities where local pride has been eroded, for example where a main source of employment (such as mining or heavy industry) has been closed down, could benefit. Being part of a community that has a large and significant district heating system would offer that community a new focus and develop other opportunities. For example, over 5,000 people each year visit the Marstal District Heating system on the Isle of Ærø. A trip to the island requires an hour's ferry ride as well as some considerable land travelling from most parts of Denmark.

There are other advantages. Almost all communities heat individual homes and domestic hot water by fossil fuel, which inevitably raises various health and safety issues. First, homes have devices that burn fossil fuel with flame. This inevitably produces some risk within the home. Secondly, there are sometimes large amounts of fuel stored in or very close to the home. This also presents a risk of conflagration. Thirdly, there are important cleaning and servicing operations regularly required to deliver heat and hot water safely. Sometimes servicing is ignored because it is too expensive or inconvenient. District Heating systems are serviced and cleaned at the plant, not in the home.

Therefore, district heating systems are simply safer. All the burning and heat generation is carried out remotely and delivered to the home through a thermal ring main. Fuel is not introduced in or close to the home. Heat is transferred by heat exchangers for the hot water supply and the radiators are connected directly to the system, a clean modern way of doing things. Cylinders are required in homes to enable hot water to be stored but most homes' existing "wet" heat exchange domestic radiators will suffice and can be connected directly to the system.

There are initial capital cost implications in building district heating systems. Payback and overall cost cannot compete with fossil fuel systems under existing financial structures. However, if we bear in mind that generally the taxpayer paid for the existing energy infrastructure many years ago, district heating systems are highly cost effective on a whole life basis.

> Instead of buying heating oil or metering gas, district heating householders pay according to the heat they use. Heat meters will calculate their charges and can also monitor the return temperatures as the thermal ring main leaves the home. If the return temperature is too high, the system managers will send an engineer to rebalance the home's radiators, increasing efficiency.

Most existing District Heating systems are built on relatively flat parts of Scandinavian countries. The solar fraction performance (the percentage of contribution made by solar to the heating system) in winter is almost nothing. This is not only due to the low light levels in winter (for example Ærø is at the same latitude as York) but also because the flat land and the proximity between collectors causes one row of collectors to cast a shadow over the next row of collectors. Mounting collectors on a south facing hillside would create a better solar fraction and better performance with less shading.

Of course, there is a great deal of pressure on land and its use has to be very carefully considered. Using sustainable energy sources, a given area of land will be able to produce a specific output of energy having regard to the energy source used and solar is simply the most efficient use of the land.

Rise District Heating System on Ærø

In the spring of 2001, a District Heating System serving the two small villages of Rise and Dunkær on the island of Ærø was commissioned. This system is based on renewable energy and is the model upon which Genersys District Heating Systems are based. The Rise District Heating System provides space heat and hot water to 115 homes. Before the plant was commissioned, mainly oil boilers heated the homes.

The system has a solar collector area of about 3600 m². This produces half of the heat energy needed.

The Rise District Heating System is owned by the 115 consumers and provides energy at a price, which competes with the price of traditional individual heating systems.

In addition to the 3600 m² flat plate solar collectors developed to produce efficient use of light at cost effective prices built to last, there is a wood pellet burning a biomass boiler. Heat produced is stored in a 4,000 m³ water steel storage tank. The biomass boiler is a wood pellet boiler of 800 kW. The whole plant is unmanned and monitored remotely.

The project started as an idea to supply Rise School and home for the elderly with renewable energy. The project promoted by the local authority that proposed that it should also encompass the two neighbouring small villages. 80% of the homeowners in the villages approved the idea. They established a consumer owned company to build the plant.

Within a year, all 115 consumers were supplied with heat from the new plant. The solar collectors have produced far more than expected.

The project has proved that it is possible to establish a district heating system, based on using 50% of the annual consumption from the solar collectors and the rest produced with a wood pellet biomass boiler.

The heat loss from the heat pipe ring main was designed to be not more than 1000 mWh per year and monitoring confirms that the heat loss is probably smaller than that target figure

Statistics from Rise District Heating System on Ærø

• Solar fraction: 50%
• Fossil fuel fraction: 50% (wood pellets).
• The solar collector surface area: 3.583 m².
• The wood pellet burning boiler: 800 kW.
• Backup boiler: not necessary (If the pellet boiler breaks there is sufficient heat stored in the tank for 1 – 2 weeks, long enough to repair the boiler).
• The annual heat sold: 2,700 mWh.
• Heat loss from heat ring main: 1,000 mWh.
• Annual demand from the plant: 3,700 mWh.
• Number of domestic radiators changed: 0.

Marstal District Heating System on Ærø

Marstal have had a fossil fuelled District Heating System since 1962. In 1994 Marstal pioneered a small solar powered District Heating System with a 75 square metre test plant. By 2004 the plant had grown into a District Heating System serving 1400 consumers. The solar panels had risen to 18,365 square metres and they provide a solar fraction of around 33%.

Many different types of thermal solar equipment have been tested at Marstal to discover the right cost effective products. Tests were carried out using flat plate collectors, vacuum tubes and focussing collectors. The engineers at Marstal concluded that flat plate collectors were the most suitable and cost effective and durable for use in district heating systems.

Marstal built a further thermal storage area in the form of a covered lined pit holding 10,000 cubic metres.

• 32 km supply main pipe
• Total heat production app. 27,000 mWh
• Thermal solar production app. 7,500 mWh
• Yearly sales: 1.9 million euros
• Staff: 1 secretary, 3 assistants,1 manager, 1 manager's assistant
• 18.385 m² of solar panels (720 x 12.56 m?)
• 150 m² technical building
• 2100 m² storage tank

lifestyle by the occupants. The theoretical home should be properly insulated to high standards - the loft, the cavity walls and it should be fitted with high quality double or treble glazed windows.

The home should be heated by a high efficiency gas condensing boiler with a modern control system, possibly a new model of a combined heat and power boiler. It uses low energy lighting throughout and the occupants carefully switch off all appliances when they are not in use.

First and foremost the home should have a thermal solar system. Most of these can provide at least 70% of the home's hot water demand, although in the United Kingdom this will means virtually 100% of the hot water demand for six months of the year and less during the shorter daylight winter months. Thermal solar should be considered as a first option because it is the most mature and cost effective renewable technology.

If the house is suitable and has enough space for panels and for a store of hot water, the thermal solar panels can produce heat energy for the home by using an under floor heat system. The lower ambient heat provided by the panels provides a significant share of the heat load, reducing the fossil fuel system demand.

Another home might be off the gas network but on the electrical mains network. Here an air source heat pump can provide efficient space heating and thermal solar panels can be fitted to reduce the load on the heat pump and save electricity.

Some of the electricity needed might be produced by a small turbine; some might be gathered by PV roof tiling. Every little bit of sustainable carbon free energy will help.

Elsewhere, thermal solar might be used in conjunction with a biomass fuel for heat and electrical energy, combined with a small wind turbine and PV cells. Although everyone will wish for a home to locally produce all its energy from renewable sources, this is really only feasible for those with large amounts of money and ideally some engineering and practical skill. There are some praiseworthy people who have achieved energy independence producing all their energy needs on site, but for many years ahead this will be well beyond the reach of most.

So, in the vast majority of cases, the home still has to use fossil fuel. However, every home should not only use as little fossil fuel as possible, it should also be a source of energy generation. Renewables should work in conjunction with fossil fuelled energy. Virtually every renewable technology does this already. The problem is lack of deployment.

The fact that deployment is not a hundred percent solution is irrelevant. It is the only solution we have. By analogy we would not invest in a medicine that only cured 40% of those ill on the grounds that it did not cure 100% of those ill, where there was no other cure available. We must do what we can. Unless we make greater efforts to combine fossil fuel technology with renewable technology, the world will drift into hot temperatures, climatic extremes and ultimately poverty for the human condition.

Chapter 7

The United Nations
and energy

The United Nations was formed to prevent war and for the amicable resolutions of disputes between nations. Over the sixty years of its history it has gradually expanded its fundamental mandate to deal with other important issues one of which is perhaps even more important than war. Its members now include almost every nation in the world.

Today most members try to adhere to United Nations resolutions. Although they are worded in deliberately vague language, capable of being construed in subtly different ways, these resolutions almost have the force of legislation throughout the world, and disobedience of a resolution sometimes involves serious sanctions, including aggression against a disobedient state.

Global warming, acid rain and pollution concern the world community. Many regard these issues as the most important that humanity faces today. It therefore comes as no surprise to learn that many worldwide organisations are sufficiently concerned to try to influence people's behaviour. The difficulty that the United Nations operates under is that its resolutions are only sometimes legally enforceable, by the very nature of politics and the fact that large and powerful states can ignore these resolutions with impunity. Accordingly, the United Nations Organisation cannot always achieve sufficient consensus to make binding and enforceable laws; it has to resort to issuing Declarations.

Many of these Declarations are no more than statements of hope and good intentions or best practice that are made while the makers of those statements mostly continue in their wasteful ways.

The United Nations' environmental activities (excluding efforts to prevent the proliferation of nuclear weapons) first began in Sweden with the Draft Declaration on the Human Environment that was made by The United Nations Conference on the Human Environment, which met at Stockholm in June 1972.

It was a forward-looking declaration at the time motivated by "the need for a common outlook and for common principles to inspire and guide the peoples of the world in the preservation and enhancement of the human environment" as they rightly declared.

The Stockholm Conference proclaimed that both natural and man-made aspects are essential to well-being and to the enjoyment of basic human rights. It declared that the protection and improvement of the environment are major issues that affect the well-being of people and economic development throughout the world.

Having praised the resourcefulness and inventiveness of humankind it continued, "*We see around us growing evidence of man-made harm in many regions of the earth: dangerous levels of pollution in water, air, earth and living beings; major and undesirable disturbances to the ecological balance of the biosphere; destruction and depletion of irreplaceable resources; and gross deficiencies harmful to the physical, mental and social health of man, in the man-made environment, particularly in the living and working environment. Millions continue to live far below the minimum levels required for a decent human existence, deprived of adequate food and clothing, shelter and education, health and sanitation.*"

It pointed out that in developed countries environmental problems are generally related to industrialization and technological development. And then, in resounding language said, *"A point has been reached in history when we must shape our actions throughout the world with a more prudent care for their environmental consequences. Through ignorance or indifference we can do massive and irreversible harm to the earthly environment on which our life and well-being depend. Conversely, through fuller knowledge and wiser action, we can achieve for our posterity and ourselves a better life in an environment more in keeping with human needs and hopes. There are broad vistas for the enhancement of environmental quality and the creation of a good life. What is needed is an enthusiastic but calm state of mind and intense but orderly work. For the purpose of attaining freedom in the world of nature, man must use knowledge to build, in collaboration with nature, a better environment. To defend and improve the human environment for present and future generations has become an imperative goal for mankind – a goal to be pursued together with, and in harmony with, the established and fundamental goals of peace and of world-wide economic and social development."*

The Stockholm delegates called upon the governments and peoples of the world to exert common efforts for the preservation and improvement of the human environment for the benefit of all people and for posterity.

The Stockholm delegates formulated 16 principles that they believed important. This was the beginning of what has been, so far, an elusive attempt to state environmental principles. Although there have been many initiatives, all of them muddle other rights and issues with specific environmental ones. Accordingly, no statement by any government or organisation about environmental matters has the clarity of some key documents in the world's history such as for example the American Declaration of Independence.

The Stockholm Principles include a "fundamental right to freedom, equality and adequate conditions of life, in an environment of a quality that permits a life of dignity and well-being" and a requirement that the natural resources of the earth and samples of natural ecosystems should be safeguarded for the benefit of present and future generations. Vital renewable resources must be maintained. Wildlife and its habitat have to be conserved. The earth's non-renewable resources must not be exhausted.

We see the first signs of concern about pollution and global warming in principle 6: *"The discharge of toxic substances or of other substances and the release of heat, in such quantities or concentrations as to exceed the capacity of the environment to render them harmless, must be halted in order to ensure that serious or irreversible damage is not inflicted upon ecosystems."*

States were called upon to avoid damage to the seas and resources were to be made available to preserve and improve the environment. States were urged to adopt an integrated and coordinated approach to their development planning "so as to ensure that development is compatible with the need to protect and improve the human environment for the benefit of their population".

Stockholm 1972 represented a first step by the international community towards coming to grips with environmental issues. As with many of these international attempts to resolve these issues, Stockholm managed to incorporate into its Principles other statements about the evils of colonialism and apartheid. These institutions, albeit wicked and evil, were no more or less responsible for environmental damage than, say, capitalism, slavery, communism, socialism or democracy.

It is simply foolish to detract from the force of what should be an environmental statement by referring to other issues. It simply weakens the impact of all issues, and makes environmentalists appear emotive and irrational and simply band wagon jumpers. What should be a clear position on, say, pollution, becomes muddied and muddled. There are many people who think that, while conceding that global warming may be an issue, the environmentalist approach is grossly exaggerated. Claiming that, for example apartheid is something to do with damage to the environment is simply foolish.

Environmental principles revolve around the need to protect the geographical environment in which we live and not to harm those physical forces which operate within that environment. With the environment secure we can argue whether, say, democracy is better for humanity than autocracy, or whether people should be treated equally. Without securing our planet, the arguments become ultimately irrelevant.

The United Nations sought to extract and proclaim what they hold to be the key principles of development of resources. In June 1992, having met at Rio de Janeiro and debated the issues of how nations should develop their own resources, the United Nations proclaimed 27 Principles that became known as the Rio Declaration on Environment and Development.

In my view the key principles are:

Principle 2: States have . . ."the right to exploit their own resources . . . and the responsibility to ensure that (their) activities do not cause damage to the environment of other states".

Principle 4: "environmental protection shall constitute an integral part of the development process".

Principle 8: "States should reduce and eliminate unsustainable patterns of production and consumption".

Many of the remaining 24 principles, while encompassing important and uncontroversial points, do not really state principles that govern the exploitation of the environment, but rather make obvious statements of fact which simply serve to detract from the most important point of all – that environmental damage knows no political borders and that it is essential to prevent polluting and damaging activities. If we poison our planet we have no future. Rio does not go far enough.

Like many such declarations, Rio embodies a series of statements that have been thrashed out as acceptable to the member states, rather than genuine principles from which we can understand how we should manage the environment. Generally, the United Nations member states placed greater importance upon their own sovereignty than upon the prevention of pollution and the protection of the environment.

Unfortunately the Rio Declaration is worded in uninspiring language that has failed to concentrate on the key issue of the control of pollution. The states attending Rio adopted "Agenda 21" in order to flesh out the bones of the principles. Here, they went some way towards addressing the problems.

Agenda 21 contains many important agreements where the states assented to diverse environmental measures ranging from managing fragile ecosystems, conserving biological diversity, combating deforestation and environmentally sound management of hazardous wastes, to issues that more closely concern the scope of this work – the protection of the atmosphere.

It is outside the scope of this work to discuss more than a small part of Agenda 21 but we must mention Section 28 of it. Clearly, many of the problems of pollution and the solutions being addressed by Agenda 21 have their roots in local activities. Accordingly, the authors of Agenda 21 declared that the participation and cooperation of local authorities will be critical to its success. They rightly understood that local authorities construct, operate and maintain environmental infrastructures, oversee planning processes, establish local environmental policies and regulations, and assist in implementing environmental policies. They play a vital role in educating, mobilizing and responding to the public, in order to promote sustainable development.

In Section 28, the United Nations mandated all local authorities throughout the world to the following tasks:

By 1996 to undertake a consultative process with their populations and achieve a consensus on "a local Agenda 21" for the community;

To implement and monitor programmes which aim at ensuring that women and youth are represented in decision-making, planning and implementation processes;

To enter into a dialogue with their citizens, local organizations and private enterprises and adopt "a local Agenda 21". The process of consultation would increase household awareness of sustainable development issues;

To assess and modify local authority programmes policies, laws and regulations to achieve Agenda 21 objectives.

To develop and employ, as appropriate, strategies for use in supporting proposals for local, national, regional and international funding.

Governments were charged with requiring local authorities to meet these objectives. The United Kingdom signed up to Agenda 21, committing all its local authorities to their appointed tasks.

Unfortunately, the wording of Section 28 is not sufficiently mandatory, and there is no real mechanism for incorporating detailed rules that follow logically from Agenda 21 into United Kingdom law, short of government legislation. At this stage such legislation seems remote. Even if they were incorporated beyond general statements there is little local authorities can do without being given the money to do it, and there is little sign of that happening at present. Most local authorities do their best to comply with Agenda 21. But generally they lack leadership as well as money.

The Rio Declaration eventually led to the adoption of the United Nations Framework Convention on Climate Change, in 1992. This Convention, signed in New York on 9th May 1992, is probably the first document in which all the nations of the world have expressed their concern that concentration of greenhouse gases caused by human activity is causing deleterious climate change and a determination to protect present and future generations from climate change. It seems astonishing, having regard to the clear wording of the convention that anyone today can feel any doubt that the planet needs protection envisaged by the convention.

The preamble to the convention identified what were to become core concerns and an almost insurmountable barrier to achieving any real atmospheric protection. "*Affirming that responses to climate change should be coordinated with social and economic development in an integrated manner with a view to avoiding adverse impacts on the latter, taking into full account the legitimate priority needs of developing countries for the achievement of sustained economic growth and the eradication of poverty, recognizing that all countries, especially developing countries, need access to resources required to achieve sustainable social and economic development and that, in order for developing countries to progress towards that goal, their energy consumption will need to grow taking into account the possibilities for achieving greater energy efficiency and for controlling greenhouse gas emissions in general, including through the application of new technologies on terms which make such an application economically and socially beneficial.*"

The preamble is important because it mentions energy use specifically and opens a loophole so large that there is little point having the convention at all. It allows poor undeveloped or underdeveloped nations to use whatever energy they can find in order to eradicate poverty and enjoy economic growth. This exemption is so widely drawn that the central objective of the convention is unattainable and the exemption is drawn on the premise that economic growth is wholly beneficial.

The convention sought to stabilise the concentration of greenhouse gases in the atmosphere at a level that would prevent dangerous anthropogenic interference with the existing climate system. The nations signing the convention (virtually the whole world) adopted five guiding principles. The first was that developed countries should "take the lead" in combating climate change. Secondly, that developing countries should not have to bear a disproportionate or abnormal burden. Thirdly, precautions should be taken to minimize, prevent or mitigate the causes of climate change. Fourthly, countries should promote "sustainable" development and lastly there should be an open economic system – free trade- leading to sustainable economic growth thus making underdeveloped countries better equipped to fight climate change.

The fifth principle is simply a restatement of the core concern – undeveloped countries do not want to stay underdeveloped. Developed countries do not want to lose their prosperity. The panacea, according to the convention, was "sustainable" growth.

The convention committed states to publish greenhouse gas emission statistics and formulate and record each nation's programs to mitigate greenhouse gas emission as well as promote technologies which could reduce human greenhouse gas emissions. Sustainable development was to be promoted and international co-operation required. Every state was supposed to take climate change into account in all their policies and also to promote scientific research that might help alleviate the problem.

In addition, the developed nations committed themselves to reduce their carbon dioxide emissions by 1999 to those of 1990. However, the signatories to the convention have allowed themselves the right to withdraw from it, upon giving notice. In other words they agree to be bound by the convention unless they change their minds.

It is clear, then, that while the Convention was being signed the nations knew that it would not affect climate change processes and negotiations started about adopting newer tougher measures for the industrialized nations. After two and a half years a protocol was adopted at Kyoto on 11th December 1997.

The Kyoto Protocol set out a system of "mechanisms and compliances which would bind the nations that ratified the protocol to reducing carbon dioxide and other greenhouse gas emissions. Later, at Marrakesh, a "rule book" was agreed setting out how the mechanisms would operate. Those states that were parties to the Framework Convention on Climate Change and became parties to the Kyoto Protocol became bound to individual targets which commit the nations to reduce their greenhouse gas emissions.

Together, so the rationale for Kyoto went, if all of these targets were met there would be a total cut in greenhouse gas emissions from 1990 levels by at least 5% in the period 2008 to 2012. However, some important points should be understood. The system of targets simply affects future greenhouse gas emissions. Greenhouse gases accumulate in the atmosphere and stay there for around a hundred years so these targets do nothing to reduce greenhouse gas levels; they simply seek to achieve some kind of reduction in the rate of emission.

The individual targets are:

Country	Target (1990 – 2008/2012)
The European Union (except Malta and Cyprus), Bulgaria, Estonia, Latvia, Liechtenstein, Lithuania, Monaco, Romania, Switzerland	-8%
The United States	-7%
Canada, Hungary, Japan, Poland	-6%
Croatia	-5%
New Zealand, Russian Federation, Ukraine	0
Norway	+1%
Australia	+8%
Iceland	+10%

Not all of the targets worked towards reducing greenhouse gases from 1990 levels. Some countries undertook to keep them at 1990 levels and others not to increase them beyond a certain percentage. This compromise arose because different countries were at different stages of their economic cycles in 1990 and a country that was producing little in 1990 as a result of an economic depression should not be penalized by this, compared with a country that was at the top of its economic cycle in 1990 and has since slipped back. This is a structural difficulty that arises when applying targets related to carbon emissions.

In addition, the largest polluter, the United States, has indicated that it will not ratify Kyoto. This nation is responsible for at least a quarter of the world's carbon dioxide emissions. Furthermore the countries which are underdeveloped have no targets at all. A country the size of India or China, if it reaches the same level of development as the European Union or the United States by using fossil fuel energy in the ways that we have already used them, will make the achievement of Kyoto targets all irrelevant, by virtue of the huge additional uncontrolled emissions that will occur.

It should be understood that no member of the European Union is likely to meet its Kyoto target, with the possible exception of the United Kingdom who may meet it as a result of policy decisions to close down coal mines, years before the Framework Convention of Climate Change was signed.

The Nairobi Declaration is another important document. Ministers and others similarly charged with the duty of protecting their countries' environments drafted it in 1997. It defined the role of the United Nations Environmental Programme and strengthened it by enlarging the Programme's mandate to six key areas of environmental policy and action:

- to analyse global trends and provide policy advice and early warning of threats, promoting the best scientific capabilities;
- to develop international environmental law;
- to implement international norms and policies, and foster compliance and cooperative responses to environmental challenges;
- to strengthen its role in the UN and the Global Environment Facility;
- to facilitate cooperation and serve as a link between all players in the international environmental field;
- to provide advice to governments and institutions.

This mandate represents an advance on Rio, but a regression from what could have resulted from Stockholm. It is also a shame that UN members have not been able to give a mandate to the UN for effective policing. Without proper international policing much of what the UN decides will be ignored or abused.

At the end of May, in the year 2000, ministers with environmental responsibility and other heads of delegation met in Malmö, Sweden to review important and emerging environmental issues, and to chart a course for the future. They were deeply concerned that, despite what they wrongly (in my view) described as the many successful and continuing efforts of the international community since the Stockholm Conference, the environment and the natural resource base that supports life on Earth continue to deteriorate at an alarming rate.

They declared what they described as the major environmental challenges of the twenty-first century. I do not set out all of the described challenges because – of necessity – some of them involve formulations of words that were inserted presumably for political purposes. The important challenges according to the Malmö Conference are:

The growing trends of environmental degradation that threaten the sustainability of the planet must be arrested and reversed.

Not enough is being done; immediate employment of domestic and international resources, including development assistance, far

December 3rd 2005; over 10,000 people are demonstrating about climate change on the Streets of Montreal while the UN conference takes place.

beyond current levels is vital.

An international environmental law must be evolved and national law developed as a basis for addressing environmental threats; we must all adhere to the precautionary approach.

The richest countries are consuming far too much of our resources. Care for the environment is lagging behind economic and social development, and a rapidly growing population is placing increased pressures on the environment.

Environmental problems arise from urbanization, climate change, the freshwater crisis, the unsustainable exploitation and depletion of resources, drought and desertification, and uncontrolled deforestation, thus increasing environmental emergencies and the risk to health and the environment from hazardous chemicals and pollution. These are all issues that need to be addressed.

Technological innovations and the emergence of new resource-efficient technologies provide a source of hope and increased opportunities to avoid the environmentally destructive practices of the past. The private sector has an essential part to play in this. Environmental considerations must be part of decision-making.

The environmental perspective must be taken into account in economic policy-making at government level and in the decisions of multilateral lending and export credit agencies.

Somewhat naively, in my view, Malmö described the private sector as having "emerged as a global actor that has a significant impact on environmental trends through its investment and technology decisions".

Of course, the private sector has always been critically important and central to initiating environmental trends. It has most often been more important than governments. For example, it is the private sector that, by marketing, creates demand, such as for hardwood from the tropical rainforests or for cars, by designing and marketing cheap and affordable models. States never founded oil exploration companies and coal mining ventures when they were highly speculative, although they may have nationalised them after they were established and successful, or strategic businesses.

Generally, the private sector acts without regard to the long-term view. In a world economy where joint stock companies' shares can change hands in a moment, the owners of, say, oil companies are a fluid and potentially changing corpus. There is no logic in their directors taking a long-term view. Furthermore there is an element of unaccountability in multinational corporations whose annual profits may be more than the gross national product of a medium-sized country and who employ tens of thousands of people in key sectors.

The private sector has always been responsible for much of the pollution and the environmental damage that flows from it. As consumers, we cannot blame the corporations. After all, we buy the cars they produce and consume the electricity they generate.

What is true, however, is that the private sector can be harnessed to redress environmental wrongs. If the government can create a business environment where companies can make a profit from providing solar collectors that reduce emissions and pollution, then the profit motive will ensure their wide distribution. If the government can ensure that those products in which an individual has to invest are provided with a tax incentive, then people will vote with their wallets. In many situations, however, environmentally conscious consumers do the sums and find that although installation works in the long-term, there is simply not enough incentive in the short- to medium-term.

In this regard, governments have a crucial role in creating an enabling environment. As the Malmö conference proclaimed "The institutional and regulatory capacities of governments to interact with the private sector should be enhanced".

Certainly, companies in the private sector that are responsible for pollution should be made accountable through the application of the "polluter pays" principle, environmental performance indicators and reporting, and the establishment of a precautionary approach in investment and technology decisions. In my view, although these are important, it is equally important to create an environment in which the person cleaning up pollution profits, as well as making the polluter pay.

> The "polluter pays" principle is important, but I wonder how often the polluter really does pay more than a token towards the real cost of environmental damage. What compensation does a person receive for a spoiled view? Mining operations dirtied much of the Durham coastline, rendering it bleak and useless for amenity purposes. Although it has been largely cleaned up, no resident received compensation for being unable to enjoy the

> beach, or swim in the sea. What compensation does the
> asthma sufferer receive from motorcar users? What
> compensation does the real sufferer from pollution
> receive for his or her distress?

If we set up a system of rules that business interests can use in order to make profits, and in the course of their activities they pollute, why should their liability for pollution be limited to the cost of clean-up operations? The polluter at best pays for some clean-up, but never, ever, the whole cost, and never enough to reinstate the environment in the condition it was before it became polluted.

In late 2005, a further climate change conference was held at Montreal. After much debate and posturing the conference was hailed a success because the parties signed a statement of intent. They are now looking for ways to reduce the level of greenhouse gas emissions by 30% to 50% within a generation. The countries of the world have agreed to work together to this end but it now seems unlikely that binding targets will ever be imposed and that any control will be imposed over the huge carbon and polluting activities of China and India, who will not be penalised for their very rapid economic growth.

The 30% to 50% range is thought by most scientists to be too little. The world faces a two or three degree Celsius rise in its climate, over and above the 1 degree it has experienced since the industrial revolution, two hundred and fifty years ago. If current greenhouse gases stabilise at 60% of their present output, we might well limit the temperature rise to three degrees.

When we analyse all these treaties, conventions and protocols we find no more than a modest statement of intent. The fundamental strategy of the Kyoto Protocol, one of emission reduction targets, is doomed to failure. There are difficulties in measuring emission reductions, and no targets for two major polluting countries, India and China. The United States will not sign up to targets. If a state gives one year's notice it is no longer bound by Kyoto, even if it has ratified the treaty.

In these circumstances the international efforts are no more than statements of intent and should be recognised as such. I do not criticise international efforts for being statements as opposed to policies and measures, because that it how the international community works and can only work in that way. I do criticise it for seeking

Climate change seems to generate lots of talking and, in relation to the importance of the issues, very little action.

simply one or two solutions; targets were critical to Kyoto; now the European Union, led by the United Kingdom, are looking at carbon trading as the panacea. Ultimately what will work is direct binding legislation in all countries and people of the world being willing to make some sacrifices in order to preserve it for their children.

Chapter 8

The European Union and energy

There are twenty five countries that belong to the European Union. These make up 6% of the world's population and are all sophisticated, highly developed and industrialised countries. The energy initiatives of the EU are important for those that live within it and for those that live elsewhere. Under Kyoto, the European Union undertook to reduce its greenhouse gas emissions (compared with 1990 levels) by 8% between 2008 and 2012.

The European Union issues "directives" which member states have to enact into their local laws. The Union also attempts to influence energy policy by campaigns. In addition the European Union requires its members to measure their carbon and other greenhouse gas emissions.

The obligation to measure greenhouse emissions is coupled with an obligation to devise and publish national and community programs for limiting and reducing emissions and report on progress each year. The reporting must include provisional data on emissions of carbon monoxide, sulphur dioxide, nitrogen monoxide and volatile organic compounds emissions of carbon dioxide, methane, nitrous oxide, hydro fluorocarbons, perfluorocarbons and sulphur hexafluoride in year; greenhouse gas emissions resulting from land-use and forestry during year.

So far, the reporting has revealed that in 2002 the total greenhouse gas emissions of the EU-15 Member States were 2.9% less than those of 1990. After two consecutive years of increase, the greenhouse gas emissions of the EU-15 Member States fell slightly compared with 2001 (by an estimated 0.5%), bringing total emissions to 2.9% below 1990 emissions.

All member states except Cyprus and Malta have targets under the Kyoto Protocol, and all but Slovenia are expected to achieve their reduction targets. The European Union as a whole also has a target under Kyoto.

As early as 2000, when the EU was much smaller, its governing body, the Council of Europe, resolved that by 2010 Europe should be producing 12% of its energy from renewable sources. It established a "Campaign for Take-Off" to achieve this end. The campaign funds and encourages pilot schemes for renewable energy projects.

The Campaign for Take Off identified certain specific key sectors which should be promoted. There were a number of ambitious targets for renewable energy which they expressed as systems or installations, rather than in terms of energy produced or carbon saved.

There should be, the campaign said, one million photovoltaic systems producing electricity. There should be fifteen million square meters of thermal solar collectors. The Council of Europe describes thermal collector technology as "almost fully mature" and states that "high quality products are available, solar systems are reliable and their productivity can be guaranteed". Domestic hot water is the main application although there are also space-heating and air-conditioning applications.

The Council targets 10 MW of wind turbine generators and 10,000 MW of combined heat and power biomass installations. There should be 10 million dwellings heated by biomass: 85% of all EC housing is heated by single house systems. The number of houses that are heated traditionally by wood furnaces is declining. With modern efficient wood furnaces emissions drop dramatically and efficiency improves from 55% to around 90%. In Austria, Denmark and Finland, biomass and other wood furnaces are used to heat whole districts and communities. The cost of investing in efficient wood burning power stations is very high.

There should be 1,000 MW of biogas installations: in this case the Council hoped to harness the methane gas produced in landfill sites and by livestock. Small experimental operations are in place and, although we hope that we are wrong, the Council's projection of 15% energy derivation from this source by 2010 looks somewhat optimistic.

Finally there should be 5 million tonnes of liquid biofuels: these, such as oil from rapeseed, competing directly with the oil industry and having the advantage of being renewable annually.

The overall investment in the Campaign for Take Off would cost, it was estimated, 30 billion euros, but of this, less than one quarter was going to come from public funds. The Council itself would contribute less than one billion euros, the remaining six billion coming from the taxpayers of national governments.

The Campaign for Take-Off is unlikely to meet its global target of a 12% carbon reduction or its individual targets. In fairness to the Council, they expected that the overall cost would be around 30 billion euros, with 6 billion coming from member states and 1 billion from the Council. The remaining 27 billion euros would come from the consumers. It seems that none of these groups was willing to make the investment required for the Campaign for Take Off to become a success and achieve its targets.

The European Climate Change Programme (ECCP) first came into being in 2001. Its has analysed various environmental measures with a view to deciding the most beneficial and cost effective in helping the EU meet its Kyoto target reduction in carbon of 8% by 2008-12. As this amounts to over 336 million metric tonnes of carbon, it is an important programme. The ECCP operates under a number of important directives, which member states are obliged to enact.

The part of the EECP that has been strongest and most successful was the focus on energy demand savings, in households and in other buildings. Less successful has been the ECCP that relates to household appliances and I shall discuss these later. Another part of the policy that has really failed to deliver has been that relating to emission savings by the transport sector. The programme successfully lowered emissions by motor cars but cannot address to projected increased use of them, so there is no overall carbon gain, rather a carbon increase. An important part of the ECCP was the new emphasis on reducing methane and nitrous oxide emissions.

The most recent estimates indicate that carbon dioxide emissions in the EU in 2000 had decreased by only 0.5% compared with 1990 levels but overall, the effect of attempting to reduce methane and nitrous oxide emissions meant that overall greenhouse gas emissions were down by about 3.8% below the 1990 levels; methane and nitrous oxide were down by 16% and 20% respectively.

In 2006 every person in the EU, according to the EU, is responsible for about 10.8 tonnes of carbon a year, whereas in 1990 the figure was 11.5 tonnes.

Why has carbon emission not reduced significantly overall? One of the main reasons was a 2.4% rise in carbon from electricity generation. In addition, as the previously less prosperous countries in this period of Ireland, Greece and Spain became significantly more prosperous; their prosperity had the effect of more consumption, more transportation and more appliances which in turn resulted in more carbon emissions. Prosperity is a good thing but only if it is managed in a way that does not harm the whole world.

It is very important to recognise the work that the EU has done in the field of reducing emissions from waste. In 1990 around 4% of the EU's emissions (the equivalent of 166 million tonnes of carbon dioxide), was created by waste, usually buried in landfill sites. The Landfill Directive aims to prevent pollution and emissions caused by land filling waste. The logic behind the directive was simple; by reducing land filling we shall reduce the adverse effects of land filling. Essentially the directive prevented member states from dumping hazardous waste and safe waste in the same landfill site. It also required pre-treatment of waste (such as sorting) to aid recycling and banned certain wastes from landfill altogether.

The rationale behind the Landfill Directive is self evident; it is a matter of shame that the United Kingdom was at least three years late in implementing it; the government explained that the reason for this was that they regarded the directive as highly technical and that as it would have serious repercussions for the waste management industry they wanted to consult as widely as possible. This is a poor excuse. The Directive is supposed to be legally binding. To consult as opposed to implement simply continues the pollution. The United Kingdom should have simply implemented it. Interestingly, most of the consultation took place after the directive should have been implemented.

The Landfill directive is now more or less implemented over the whole European Union and as a result significant emissions of methane have been prevented.

Another important Directive was promulgated – the Energy Performance of Buildings Directive (EPBD) in 2002. Its underlying philosophy is that influencing energy supply is really difficult to do, but influencing demand and in particular behavioural demand and consumption, is possible by improving energy efficiency.

As EU households are responsible for about a third of the EU's carbon emissions, the directive requires member states to create laws

Leachate from a landfill site in Cardiff, Glamorgan, South Wales.

which will have the effect of making households use energy more efficiently. Nine years before this directive was passed, the EU issued another directive on the energy certification of buildings. The EPBD builds on this by requiring a common methodology for calculating the integrated energy performance of buildings; minimum standards on the energy performance of new buildings and existing buildings that are subject to major renovation; energy certification of new and existing buildings and, for public buildings, prominent display of this certification and other relevant information. Certificates must be less than five years old; regular inspection of boilers and central air-conditioning systems in buildings and assessments of heating installations in which the boilers are more than 15 years old.

Although the EPBD was supposed to be fully implemented by January 2006, the UK had none of it implemented by then. Part of the EBPD which advised designers to "consider" low and zero carbon technologies (rather than mandating them) resulted in changes to Part L of the Building Regulations (which included the mandating of gas condensing boilers), with effect from (mostly) April 2007

Again the United Kingdom's poor performance in implementing this directive on time is a matter of concern. While the UK government prevaricates, the environment continues to suffer.

Energy performance of buildings: here high quality insulation is being bricked in at a new building being constructed in Sodertalje Storgaton, Sweden.

It is expected that by June 2007 a "Home Energy Report" will become mandatory for UK house sales as part of a buyer's information pack but the UK's other obligations – for boiler inspections, rental housing, non domestic building certification, public display of energy certificates, air conditioning inspections, qualifications for inspectors and accreditation of them all was unknown by January 2006 when the directive was supposed to be in force. It is very difficult to see any commitment from the UK to alleviating climate change when it cannot even comply with a European directive that is not onerous and not expensive.

Rightly, the EU identified that the largest potential savings in carbon emissions can be made by improved energy efficiency in households. The first thing was to increase energy and carbon awareness, provide information and label products and energy using appliances so that consumers could see how much energy they consumed.

The European Community implemented a framework for energy labelling of household appliances. At the time of writing, refrigerators, freezers, fridge-freezers, washing machines, tumble dryers, dishwashers, lamps, ovens and air conditioners are required to have energy labels so that at the time of purchase the buyer can understand what the energy cost of the appliance will be. The EU has been very slow to expand the numbers of appliances that need energy labels. Only ovens and air conditioners have been added to the list of mandatory labelling since 1998.

Now, more than ever, it is important that people can see how much energy an item will use before they buy it. It may well be cheaper to buy a more expensive product that uses less energy than a cheap product that uses a lot of energy; low energy light bulbs are a case in point, which over their lifetime work out about seven times cheaper than their tungsten equivalents.

But not everyone does the calculations when they see a label and however much information is on a label it simply provides an indication. It is far more useful to employ minimum efficiency performance standards for domestic appliances. Provided that an appliance still does its job, why should it not be designed to use less energy?

So far the European Community has required minimum efficiency performance standards in only three types of products: cooling appliances - refrigerators, freezers, fridge-freezers, ballasts for fluorescent lighting (only to be fully implemented in 2008) and hot water boilers.

By contrast, the United States of America, that country that abhors regulation of businesses and is considered to be energy greedy by some, has minimum efficiency performance standards that apply to 28 different products in all 50 states. The products that exceed the standards are given the prestigious "Energy Star" rating and the standards are improved every five or six years so that manufacturers can design the new standards into the product and consumers, when they replace their appliances, will do so with a much more efficient machine.

The US system has involved expenditure by the federal government's Department of Energy of over 200 million dollars over the past 20 years. That money has been very well spent; consumers have on average saved $75 for every $1 spent by the program.

Australia, when it decided to implement standards, simply chose what it believed to be the best standards from around the world. It was a far cheaper way to achieve high quality standards; they chose many American standards but also standards from Canada and Malaysia. It is very difficult to understand why the European Union does not make faster progress on minimum efficiency performance standards. Increasingly, Europeans are using more domestic appliances powered by electricity. As we have seen, electricity produces oppressive amounts of carbon and anything that saves electricity without damaging the performance of the product should be a "gimmee".

In 2003 the European Union, by directive, operated a carbon trading scheme. Under this scheme, individual businesses in the fields of energy, iron and steel production and processing, mineral industries, paper pulp and card industries would be allowed to emit so many tonnes of carbon dioxide (or equivalent greenhouse gas) each year. It is called the Greenhouse Gas Allowance Trading Scheme.

These industries have to apply for a permit to emit greenhouse gases. The permit, for the first few years of the scheme (from 2005) allows specific emissions (up to 95% of the requirement) virtually without cost to the specific industry. If the business exceeds its quota it will have to pay a penalty fine. The fines can be avoided by "purchasing" carbon savings from participants that have managed to emit less carbon than their permits allowed.

The scheme is still in its design stage. Each Member State must draw up a national plan complying specific criteria in the Directive, indicating the allowances it intends to allocate for the relevant period and how it proposes to allocate them to each installation. A number of member states have submitted plans and the European Commission has assessed the national allowance plans of Denmark, Germany, Ireland, the Netherlands, Austria, Slovenia, Sweden and the United Kingdom. None of these plans has been rejected outright, but certain aspects of the plans of Germany, Ireland, Austria and the United Kingdom have been rejected.

> These eight plans represent almost half the overall estimated volume of allowances for the first trading period (2005-2007) - a total of more than 2.88 billion tonnes. Denmark's plan covered 100.5 million tonnes, Germany's 1, 499 million tonnes, Ireland's 66.96 million tonnes, Holland's 285.9 million tonnes. The Austrian plan covered 98.24 million tonnes, Slovenia's 26.3 million tonnes and Sweden's 68.7 million tonnes. The United Kingdom's plan covered 736 million tonnes.

If the scheme is a success, the European Union will build on it. I personally have some doubt about trading carbon allowances. It seems to me that licensing amounts of carbon output is sensible but allowing operators to "buy up" un-emitted output is artificial. It does provide industry flexibility but do we have time to allow this? If carbon emission is harmful then a process of gradually forcing its reduction industry by industry, in the same way that the USA has gradually required domestic appliance efficiency, works better without the intervention of what is in effect a poison trading scheme.

The European Union is also looking at a scheme for carbon trading based on energy efficiency – insulation. Here the concept is to create white certificates that can be traded like carbon allowances.

In the case of carbon dioxide there is no end use for it. The actual emission savings cannot accurately be defined; a project may make greater or fewer carbon savings than estimated and much of the carbon saving itself, even with the very best technology, ultimately depends on human behaviour.

Carbon emitters will ultimately have to be trusted to some extent in terms of the savings that they say they are making. It will be almost impossible to police such a scheme. In addition the inability to define actual and precise savings makes any carbon trading scheme open to abuse. In these circumstances it may well be that very little actual benefit can come from a carbon trading scheme.

Saving carbon emission cannot be undertaken by creating an artificial market in carbon dioxide emissions.

In the case of the proposals on white certificates this is in my view a complete waste of time and deeply flawed. In many EU member states, the UK for example, much insulation is installed by legislative mandate by the energy companies. Why should they be further rewarded for doing what they are already being obliged to do and which they do at the expense of the public? What end use is there for White Certificates? Every market has to be founded on an actual end use for the commodity in the market. Experience of market mechanisms shows that often trading leads to what is in effect gambling. More commodities are bought and sold on the futures markets than are ever used. Can we really expect a benefit to the planet from such a scheme?

There are three areas where the EU has failed to develop a coherent strategy. Agriculture and forestry is the first area. This sector is responsible for around 10% of the greenhouse gas output. The EU has significant potential to use locally grown biomass and biofuels instead of fossil fuels. The benefits and disadvantages of biomass are not completely understood. For example if we burn trees for fuel we generate carbon, which we can remove from the atmosphere by growing more trees. Trees, however, are an important short and medium term store of sequestrated carbon. The balance between cutting or thinning them and leaving them, in net carbon terms, still needs research, which the EU is sponsoring.

The second area of inactivity is aviation fuel. This accounts for 3.5% of the EU's greenhouse gas emissions. Aviation fuel is not taxable, under international treaties. Aircraft take many years to develop and most aircraft flying today where designed when aviation fuel was cheap and plentiful.

Thirdly, shipping fuel accounts for around 2.5% of the EU's carbon output. Again, this is an area of inactivity, for much the same reasons as aviation fuel.

Finally the European Union does, as can be seen, have an environmental policy. It is flawed and it is developing too slowly but the overall policy has some coherence and a real direction. However, the EU does not have an energy policy. This means that the EU's environmental rules inevitably come into conflict with individual states' energy policies and individual states are often slow to adopt environmental policy because of the fear of prejudicing their own industries and businesses. If the European Union adopted a coherent energy policy instead of the present free-for-all that now exists, implementing environmental policy should be quicker, easier and more acceptable to the individual nations.

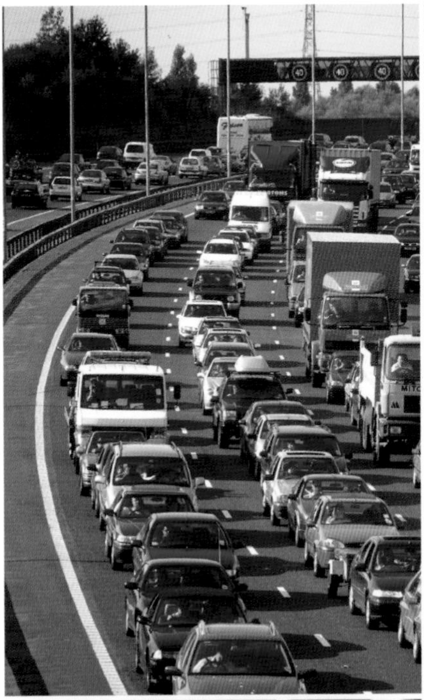
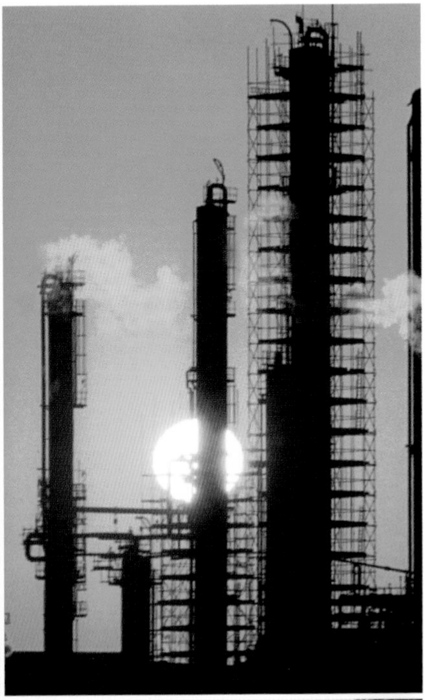
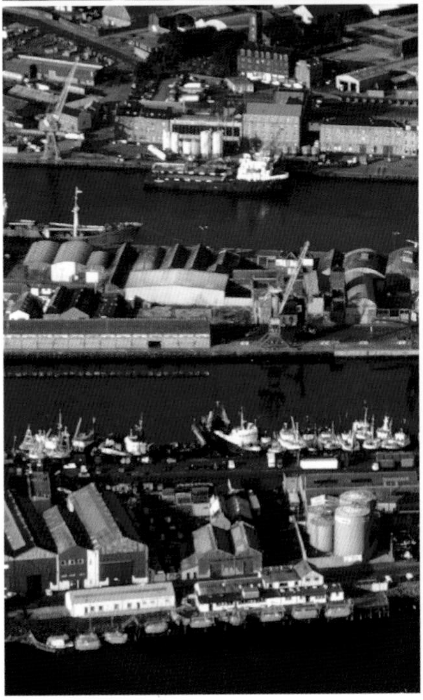
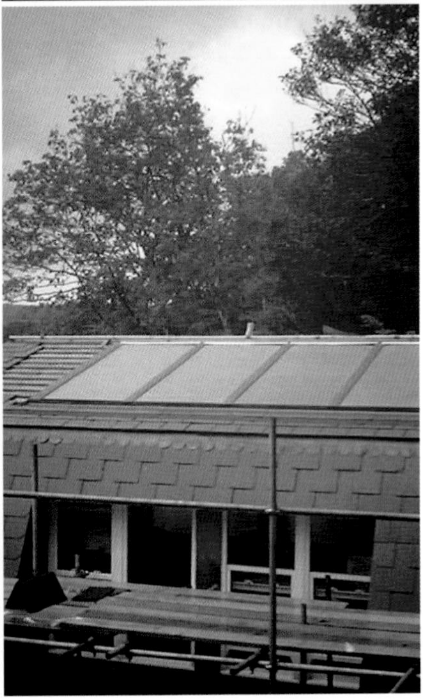

Chapter 9

The United Kingdom's energy policy

The industrial landscape of Port Talbot, South Wales where an oil refinery, steelworks and power stations are located.

The United Kingdom is a member of the United Nations and an important constituent part of the European Union. All European Union countries are obliged by treaty to adhere to certain obligations. These treaties oblige a reduction in greenhouse gas levels. Membership of the European Union requires certain specific energy obligations, which we shall examine later. However, in addition to these binding duties, the United Kingdom has been examining what else should be done and how it should be done.

In February 2003, the Government published a new energy policy, in a white paper, which claimed to create a change in emphasis. It subtitled the policy "Our energy future – creating a low carbon economy". The new policy was important. The Prime Minister himself, in a foreword, pointed out that climate change threatens "major consequences": that our energy would become increasingly dependent upon imported fuel whilst we faced the challenge of keeping energy affordable for industry and for people. Mr Blair has correctly identified the challenges and problems.

The new policy, Mr Blair said, would tackle these challenges. He also said that the United Kingdom was showing leadership; we were as a nation on the path to a 60% reduction in our carbon emissions by 2050. The United Kingdom has a legally binding treaty obligation to reduce greenhouse gas emissions by 12.5% to 1990 levels by 2008-2012. The Government has also set a separate non-binding target of reducing these emissions by 20% from 1990 levels by the year 2010. It wants to make good progress towards that target.

The 12.5% target will probably be met but not as a result of any policy of the present government. The meeting of the target will be due to the fact that in the 1990s far more power was generated by coal than now. The closing of the mines and the decimation of the UK coal industry led to a "dash for gas" among power station operators, and as we have seen, gas produces fewer carbon emissions than coal. Ironically, achieving the 12.5% target is as a result of Mrs Thatcher's and Mr Major's policies, and nothing to do with Mr Blair's policies.

UK Government policy, created to tackle problems in energy and climate change, can only ever have a limited success. Climate change is a problem that needs to be addressed by the whole world. However, a single nation can show leadership and a nation that is genuinely tackling climate change will inspire other nations and show them how matters can be improved and those other nations will implement effective climate change policies. Conversely, if an important nation fails to show leadership in tackling climate change then very few other nations will be concerned to implement climate change policies.

The other problems of energy supply, reliability and affordability cannot be considered by one nation in isolation, unless that nation has huge reserves of fossil fuel. Most nations, and certainly all of the developed nations and the developing nations, compete with each other to buy and source energy. In a market, those will-

The M25 motorway near Egham gives some idea of the volume of UK traffic that exists on what is a very small island.

ing to pay the highest price will secure the product if it becomes scarce. Affordability and supply are dependent upon world markets, and can only be marginally affected by policy.

Nevertheless, a good, well implemented energy policy can help tackle climate change and the other challenges about which Mr Blair wrote. I shall examine what the United Kingdom's energy policy consist of, how it is implemented and whether we, as a nation, are showing leadership and inspiration.

The policy is founded on four cornerstones: the "decoupling" of economic growth from energy use and pollution; a secure energy supply; an affordable energy supply, and to eradicate fuel poverty. I shall look at them each in turn.

The first cornerstone of energy policy – decoupling economic growth from energy use and pollution

Economic growth has always required increasing levels of energy use. In turn, the effect of economic growth is to stimulate and increase energy use. The more energy we use the more pollution and atmospheric carbon we create. If we stop using energy it is feared that we may lose economic prosperity to nations that continue to use it. The UK policy aims to (i) reduce the energy we consume by promoting energy efficiency and (ii) use no carbon or low carbon energy sources more, so that some significant energy demand is provided by renewables.

It follows then that half of the constituent of the first cornerstone of "decoupling" is *energy efficiency*. Energy efficiency is important. Energy is wasted by heating poorly insulated buildings, by uncontrolled use of lighting and by using household appliances that use more energy than they need to. Energy is also wasted by inefficient transport and in some cases by inefficient industrial processes.

The Government recognises that the costs of becoming energy efficient may in certain cases be uneconomic. Some households simply do not have the capital available in order to make energy savings and are doomed to having to pay for expensive, inefficient energy. Clearly, people on low incomes and people of limited means are particularly helped by energy efficiency because they can live warmer, more comfortable lives within their means.

Over the past 30 years or so the energy intensity, that is to say the ratio of energy consumption to gross domestic product has improved by 1.8% a year. By improving efficiency standards in households the

The Baglan Bay chemical works gently polluting the air at sunset.

government expects to save half of the carbon that they want to save. They have introduced higher standards for building new homes, including obligations that new boilers fitted should be the more efficient condensing type, rather than the less efficient non condensing type. They have introduced rules requiring, by a system of Standard Assessment Performance, new buildings to be built with better insulation, and less wasteful draughts, and are trying to introduce regulations to make domestic appliances more energy efficient.

The key measure, however, is the Energy Efficiency Commitment, (EEC). EEC was already in place before the White Paper, although the government did subsequently increase the EEC system. Under EEC, every energy supplier is obliged to meet a target (based upon how many customers they have) for energy saving. They can at present virtually only meet their targets by spending money to subsidise insulation measures (loft and cavity wall insulation) and by installing low energy light bulbs and by installing condensing boilers.

There is a very limited provision in the EEC scheme for the installation of renewable energy, but the EEC scheme is administered by the Office of Gas and Electricity Markets (Ofgem), and the structure of the scheme together with the way it is being administered makes installation of renewables virtually without incentive for the energy companies, who will continue to concentrate on insulation measures until 2008. So far over 98% of all EEC expenditure has been on insulation. The energy company must apply half its measures on poor and vulnerable households.

> The EEC scheme is mandatory and if an energy company fails to meet its target it will be fined. The fine could be as high as 10% of the turnover of the company, so meeting their targets, expressed in terms of measures rather than carbon saving, becomes critical to them. Ofgem, who spend about £1 million a year in administering EEC, estimate that 3% of every household bill is applied towards EEC expenditure, and this percentage is expected to rise over the next few years. It is estimated that EEC spending by the energy companies amounted to about £400 million in 2004, with all of the money raised coming from the pockets of their customers.

The fact that the targets are expressed in measures to which approximate carbon savings are applied, rather than in expenditure or actual carbon savings, and the way that the scheme is actually an obligation imposed on businesses that is in direct conflict with their own core business of selling as much energy as possible, seems illogical. I suspect that the energy savings by insulation have been exaggerated.

I think that the EEC scheme fails to recognise one important factor. When energy companies try to sell measures, mainly loft and cavity wall insulation, they are regarded by a section of the public, used to illusory "special offers" and "final reductions" as not offering a genuinely discounted deal. I think that there would be greater public take up of insulation measures if they were undertaken by specialist insulation companies backed by tax credits given to those taking up the insulation, rather than by involving the energy companies and Ofgem. I think this is a structural failure which arises where measures are being carried out by those whose primary interests are served in not carrying out the measures.

The second part of the cornerstone of decoupling is promoting an increased use of renewables. For the government this really means wind turbines. Legally, electricity generating companies have an obligation to generate a percentage of their electricity from renewables sources. This is known as ROC – the renewable obligation certificate, which is a means of rewarding generators with a "carbon bond" which can be traded between themselves, if one fails to meet is obligation to generate by renewables. This is also administered by Ofgem.

> In the same way that EEC has become a virtual insulation business, ignoring all other efficiencies for practical purposes, so ROCs have created a position where wind turbine generation has become virtually the sole means of fulfilling the Renewables Obligation. The system in effect depends on taxpayer subsidy – around £750 million to £1,000 million a year. 70% of the renewable wind generators' income comes from taxpayer's money. Are the wind generators in the energy business or the subsidy business?

The renewable electricity target for 2010 has been set to 10% of generated electricity. This amounts to 33.6 TWh of electricity. If this target is met 9.2 million tonnes of carbon dioxide will be saved; this amounts to only 1.7% of UK carbon emissions and only 0.0004% of global emissions. The National Audit Office has commented that as a means of reducing carbon dioxide, the renewable obligation "is several times more expensive than other measures. There is no target for renewably generated heat and hot water.

Of course, generating electricity by renewables will always be very expensive. Equivalent energy can be generated for heat at a fraction of the cost. That is why thermal solar installations are more popular with the public as household measures than any other renewable measure.

I do not criticise expenditure on wind turbines; I do find it illogical that other forms of generation of energy are not given the same priority or subsidy that wind receives. There is no renewable heat obligation, even though heat energy accounts for nearly half of our energy expenditure and, for most homes, the heating bill is greater than the electricity bill. Offering a credit for renewable electricity generation while not offering a credit for heat generation simply makes no sense. It is as though the Government equates energy with electricity.

In 2002 the government announced a system of subsidies for small scale renewable energy generating systems. This was a scheme administered by the Energy Savings Trust for only photovoltaic systems. This initially allowed for grants of up to 50% for PV installations. The funding was very small – probably less than £26 million. By September 2005 1,400 projects had been funded leading to a saving of an estimated 20,000 tonnes of carbon dioxide.

In early 2003 the government announced the "Clear-Skies" program, another set of subsidies to be shared between all the other renewable technologies. The announcement preceded the availability of the grants by five months, so it virtually brought to a halt all sales on these other renewable technologies for five months. The funding was tiny – when the scheme ended in March 2006 less than £12.5 million of tax payers' money had been spent in grants on all these other renewable technologies, including viable and cost effective domestic solar thermal systems.

It is extremely difficult to understand why photovoltaics, being so inefficient and expensive, merited not only twice the pool of money that applied to all the other technologies but also much larger individual grants. A person buying a solar thermal system would get £400 out of an overall cost of around £3,500. Another person wanting to install a PV system (that would save around the same amount of carbon dioxide emissions) would pay £10,000 but get a subsidy of £5,000. It made no sense.

There was a great deal of hope raised in 2003 when this policy was published, that the new building regulations would require some form of local energy generation from each new home. This never materialised. The building regulations are administered by the Office of Deputy Prime Minister, not by DEFRA or the DTI, and although they are of prime importance in terms of designing an energy policy of the future the OPDM seems to have been unable to seize the opportunity to make a positive change. It is an opportunity delayed to the detriment of the environment. Fitting, for example, thermal solar panels to homes that already exist is about twice as expensive as fitting them to a home in the course of construction. It is very difficult to understand why mandatory microgeneration installation in all new homes does not exist.

When we analyse the composition of the "decoupling" cornerstone, we find nothing more imaginative than subsidising the insulation of homes and building wind turbines. A cornerstone of energy policy needs to be sound, robust and composed of many different elements, as well as being properly funded. The cornerstone we are told is made up of energy efficiency and renewables but in truth this means only insulation and large scale wind turbines.

Aberdeen is the United Kingdom's largest oil city. Much of the docks are used to service and supply the offshore oil industry.

We have to recognise that the amounts spent are very modest by any standards. Given that climate change is such a problem, should we not devote much more of our resources to combating it? The whole sum of what the Clear-Skies scheme spent each year was less than the amount of public money that was spent by 659 members of Parliament, who managed to spend £7.1 million in one year on travelling expenses. The Photovoltaic Demonstration Scheme provided over three years less money that the taxpayer spent is in subsiding Members of Parliament's homes over a like period. The subsidy for wind turbines is less than a third of the annual money raised by the taxpayer to fund the BBC.

The second cornerstone of energy policy: security of supply

For many years the security of the nation's energy was the main feature of energy policy. To achieve this we need a diverse system of supplying energy using many different fuel sources and renewables. As we have seen the United Kingdom is producing less fuel. Most coal mines have been shut and the oil and gas from the North Sea is running out.

2006 sees the United Kingdom as a net importer of natural gas

Genersys solar system at social housing in the London Borough of Southwark. This sophisticated installation works with in conjunction modulating gas condensing combination boilers and was installed in 2001.

mal is environmentally benign, producing no carbon. The annual carbon savings alone would be over 12 million tonnes. Solar thermal would continue to produce local energy regardless of foreign supplies, wars and natural disasters. It is not a 100% solution, but making even a small part of our energy demand secure would be a significant achievement.

There are other important strategies that can significantly improve the plant margin. For example, homes using electricity for heat can use heat pumps to reduce the electrical requirement. Small wind turbines can also help, as can local cogeneration, which saves energy by reducing the transmission losses. Very little of these techniques are being used. At the moment, the second cornerstone of security is too fragile to provide any support.

The Third Cornerstone of energy policy: affordability of energy
The energy sector costs around 4% of our gross domestic product. In the 1980s the UK began to denationalise the energy companies and now there is real competition among them in a sector that has been rationalised and operates efficiently. This had the effect of driving down prices so that by 1992 energy prices fell in real terms for most consumers, until they became amongst the lowest in the European Union.

However, the reduction of energy prices occurred against the background of falling or stable fuel prices. In the last year or two fuel prices have all risen enormously. The basis of the fuel price rises has been the increase in oil prices which in turn has dragged natural gas pricing higher. Now energy prices are rising quickly in real terms throughout the world. It is very difficult to see what the government can do to prevent these very high rises – in excess of 30% a year - under the present privatised energy distribution system.

The government does tax energy. Most household bills pay Value Added Tax at 5% on their energy; renewables are treated equally with a 5% tax on the installation. The greater the energy cost the more revenue the government raises.

The Government introduced a Climate Change Levy in April 2001. The tax applies to natural gas, electricity, oil, coal and all similar fuels (except fuels used in road vehicles) which business, commerce and industry use for heat, light and power. Non-business use of energy or energy used by charities is not taxed. Additionally, public transport is exempt as is renewably generated electricity.

The tax is calculated by applying a rate to a theoretical nominal unit of energy. The Government decided that it is virtually impossible to relate the Climate Change Levy to the amounts of carbon dioxide emitted by each form of energy used. Using the energy content of fuels is thought to be simpler and would not encourage industry to switch from one fuel to another because of tax reasons. The levy is equivalent to a 10% surcharge on the electricity bill and a 15% surcharge on the gas bill.

Certain business uses of energy – such as the energy used to propel trains – do not attract the levy. Certain specific industries – cement, brewers, motor manufacturers, steel makers and renderers to name but a few – have special agreements and special discounts. These seem to apply regardless of the carbon they create. We have already seen the vast amounts of carbon dioxide the cement industry emits yet it has exemptions from what is presented as a carbon tax because it has met "targets" to reduce carbon dioxide emissions. Nevertheless the cement industry is still responsible for around 2% o the UK's emissions and is not paying for the effect of the carbon it emits.

A business pays tax on its profits. These are calculated by deducting most expenses from its earnings. Some costs are not deducted in the normal way. If a business invests in machinery that may last twenty years it is not usually possible to deduct the cost of the

I have tried to analyse the energy policy of the United Kingdom by using the internal criteria of the policy itself and the way in which the policy was presented by Mr Blair. I must admit that I find the policy confusing; there are many excellent parts and important sentiments but as a cohesive whole I think the policy does not make sense. I think this is revealed when an analysis is attempted.

Using the analogy of cornerstones I find that two of the cornerstones – those of affordability and fuel poverty eradication are really supporting just one corner. The third cornerstone – of security – has no substance at all and supports no part of the structure. The cornerstone of "decoupling" is flimsy and supports very little. The policy is unlikely to succeed in any of its objectives. Overall, the policy shows no leadership. No country in the world will look at the UK's energy policy as a model.

It is very easy to criticise a government policy and much harder to offer a viable solution. Governments do not consider energy in isolation. They have to get re-elected in democracies which often means that they can only do what is politically safe and take a short term view. It came as no surprise, therefore, when the Government announced a further review of energy policy in late 2005. It seems that now nuclear energy is back on the agenda – a clear indication that the security of supply "cornerstone" was false.

Energy policy needs a long term view. More importantly energy policy has to be founded upon principles, rather than made up of a collection of altruisms. I have suggested four principles in the first chapter – benign energy first, no unnecessary use of energy, the conservation of energy and the need for the polluter to pay. If we adopted these principles and worked out an energy policy around them we would have a chance of getting this critical part of government right. We would need to give the Energy Minister the power to over ride his colleagues – including the Chancellor of the Exchequer, otherwise the policy will fail. I have no problem with this because I regard energy policy to be the most important part of the government's duties.

Every means of creating energy must be considered in forming a new policy. Energy particularly that used in the home and by cars will have to become more expensive where malignant carbon emitting sources are used. Low and zero carbon energy generation must be encouraged and rewarded. Unpopular decisions have to be made if we are to avoid our world in the future being a substantially worse place in which to live. Expensive decisions also have to be made and paid for now. They have to be paid for by the community. There can be no free lunch.

The target of reducing our carbon dioxide emissions to 60% from 2003 levels certainly cannot be attained by the policy set out in 2003. In March 2005, an important study by the Environmental Change Institute was completed. The study analysed whether the UK residential sector could deliver the carbon savings by 2050. It concluded that it could but there would have to be a dedicated effort instituted now.

The housing stock must be improved and energy inefficient houses would have to be pulled down and replaced with efficient homes. All cavity walls and lofts have to be insulated. Lights and appliances that are not energy efficient have to be banned. Low and zero carbon technologies must be installed as a matter of course. Every new home must have at least two LZC technologies and they should also be fitted to existing homes as a matter of course.

As we have seen the Government missed an important opportunity in house building by not requiring even one form of microgeneration, let alone two forms, when they revised the building regulations in 2005. Home insulation to save energy is being promoted to the public by the very businesses that sell energy. It is not being required by law. Lighting is still a matter of choice. Tungsten lighting is still the norm instead of lower energy using lighting. Appliances are still being sold which use far more electricity than they need to – digital television set top boxes constantly draw power.

All of these errors can be remedied by legislation. Of course, there is a cost, and that cost may not be popular. But the price of not doing these simple things will be far higher than the short term cost and if we can find the resolve to legislate within a framework of ethical guiding principles, that may save us from being cursed by future generations who live where we live now.

Chapter 10

The United States of America – hero or villain?

Smog over New York City; the city uses huge amounts of energy every day and occasionally suffers from thermal inversion trapping smog.

4% of the World's population emits at least 25% of the world's human carbon dioxide emissions. This startling fact is often used to argue that the US is energy greedy and uncaring about the effect of its lifestyle on the rest of humanity. After all, as many people point out, the United States refuses to ratify Kyoto Protocol on Climate Change and its leaders seem to have an unhealthy relationship with sections of the oil industry. The inhabitants mainly drive around in oversized overpowered cars that pollute excessively and live in overheated or over air-conditioned homes. At leisure they drive power boats and fly all over their country. That is perhaps a typical European view.

The United States of America pollutes by itself more than several continents combined. Its insatiable appetite for energy causes it to manufacture one quarter of all pollution in the world for the benefit of its inhabitants who constitute only a tiny fraction of the world's population. If you add the pollution caused by the energy that goes into making the products consumed in the United States, the overall emissions created by or on behalf of the United States may well constitute one third of the world's pollution.

Every country, and in this the United States is no exception, wants to preserve and enhance the living standards of its people. Protecting the people and their interests and welfare is the point of any form of government; if a state does not do this the rationale for the submission of people to their government's authority does not exist. This has always been the case. People submitted to government by a king because the king protected them. Similarly, people submit to democratic government because it also protects them and cares for them.

From the point of view of the United States, Kyoto threatened its way of life because if the US submitted to targets when other major polluters did not, it would impact drastically on the US economy; it would have to make real sacrifices while its competitors, China and India in particular, would be allowed to emit carbon at increasing levels and a result prosperity would move from the USA to China and India. Turkeys do not vote for Christmas.

Energy is central to economic prosperity. It can provide all our needs as well as all of our luxuries. It can enhance our pleasures and prolong our lives in a healthy manner.

If American industry were forced to make big reductions in CO_2 emission this would undoubtedly be extremely damaging to the US economy. Over the two hundred and thirty years since America was established as a nation, the American people, comprising in the main the descendants of Europeans, Africans and Asians, worked hard and mostly indiscriminately to become prosperous. That pursuit of prosperity has been a remarkable feature of America and in many ways peculiar to it. In very few other countries, even today, can a poor person have a realistic opportunity of becoming a rich person.

The ability to become wealthy in America has given it an approach to economic activity that has so far been unrivalled. Its land mass lies at the same latitude as northern China and much of Mongolia; it has become highly populated and economically the most successful country in the world. Much of the landscape has

The Trans-Alaskan oil pipeline winding its way through the countryside bringing oil from the northern most state in the Union.

been changed and many of its ecosystems have been irretrievably altered. People live well in frozen winters and desert summers, all by the exploitation of energy. Cheap energy also gives Americans an ability to travel extensively without much regard to the cost, and to consume in many cases to their hearts' desires.

Many Americans have concerns that this high level of consumption will drastically drop if energy is expensive. This in turn will affect the whole economy causing recession and depression. People will no longer be able to live the American dream. These are real concerns. Therefore it is not surprising that Americans often feel that the world is asking them to make sacrifices that many other nations are not making. They are right in this view. Very expensive energy will change the American way of life, as they know it.

However, expensive energy is a prerequisite for environmental safety. As we have seen, we need to reduce carbon and other emissions and reducing them is not a cheap process. If a country like the United States is to reduce emissions very significantly then a large proportion of its economic resources will have to be applied in this.

In the course of the arguments over the environment, opposing sides often become entrenched and forced into holding positions that are not logical. There are people in the United States who claim that global warming is not a threat and there are people in high authority outside the United States who claim that the United States is callous about the needs of the rest of the world. Neither view is accurate.

American scientists are generally more cautious about the impact of climate change than scientists in other countries. Some European researchers are predicting big global temperature increases by 2100, but there is a more conservative view of climate change among scientists in the USA. Many American scientists tend to think that the European predictions are greatly exaggerated but they still believe that there is a global warming problem and that it is serious problem.

For example, William Dickinson of the University of Arizona argues that global warming is going to be at the lower part of the range predicted by the Intergovernmental Panel on Climate Change. He points out that the amount of carbon dioxide in the atmosphere has increased from pre-industrial levels of about 280 ppmv to the current level of about 381 ppmv. The calculated temperature rise due to the direct effect of CO_2 itself is about 0.5°C, only marginally less than 0.6°C over the past century.

If carbon dioxide in the atmosphere doubles or trebles (compared with levels before the industrial revolution), there would probably be a rise in temperature by 2100 of between 0.8°C–1.6°C. For the planet to get hotter, then the carbon dioxide levels need to trigger an increase in water vapour, argues Dickinson. The truth is that no one knows for certain how the triggering effect of carbon dioxide on water vapour works. Either it works intensively, in which case reductions of small carbon emissions are all that is required, or it works less intensively, in which case very large emission cuts are needed. Dickinson is pointing out an area where we do not know what will happen, we can only surmise.

It is not possible to dismiss the theories of American scientists like Professor Dickinson as merely serving American interests and many European commentators do a disservice in characterising them as such. They are respectable and whether they be right or wrong are no less respectable for being not entirely in accordance with the opinions of others.

Against this background, many readers may think that there is no emission control or other anti-global warming activities in the United States. There is in fact as much, if not more positive activity being taken in the United States, as there is in the European Union.

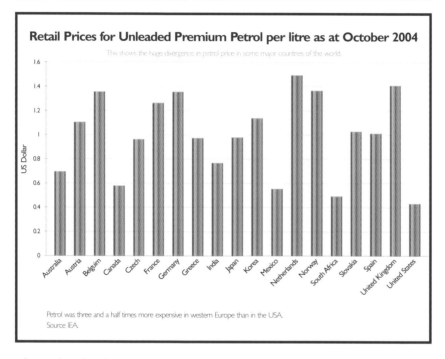

Retail Prices for Unleaded Premium Petrol per litre as at October 2004

This shows the huge divergence in petrol price in some major countries of the world.

Petrol was three and a half times more expensive in western Europe than in the USA.
Source IEA.

The US has the cheapest petrol (gasoline) in the world; these prices are based on a litre of unleaded fuel in October 2004. The price of petrol is very relevant to its use, particularly its use in leisure activities.

In the US control is based on a "carrot and stick" approach.

Some American policymakers such as Eileen Claussen, president of the Pew Centre on Global Climate Change and a former US assistant secretary of state, has indicated that the United States would not be forcing carbon emission reductions but approaching the problem by a mixtures of incentives and penalties.

America still counts its energy in British Thermal Units, rather than Kilowatts. (You can convert the measurements using the information in the glossary.) America presently consumes around 100 quadrillion Btu each year, producing around 70% of its consumption. 39% comes from oil, 23 from coal, 23% from natural gas and 8% from nuclear power. The rest comes from renewable sustainable sources. Around 3% comes from hydro production but over 3% from sustainable biomass geothermal and wind. In 2003 America consumed 84.34 quadrillion Btu of fossil fuels, 7.97 quadrillion of nuclear and 6.15 quadrillion Btu of renewable energy.

In 2002 US generators produced 758 GW from fossil
fired stations, 105GW from nuclear fired stations and 95
GW from hydro electric stations. These are large
numbers. Germany, for example, produced 114 GW from
a combination of all these sources. If you then calculate
the electricity generated from renewables (not including
hydro electricity) you will find that while Germany
generated 12 GW, the United States generate 20 GW.

The US government, through its various agencies, sagely warns
about acid rain and global warming. There seems to be some ten-
sion between the US environment agency and its view of the seriousness of global warming and the views of the executive – the President.

It is accepted that the United States is a vast consumer of electricity and that its percentage of power generated from renewables
is low but nevertheless generating 20GW in 2002 is significant and
does not lie easy with the theory that the United States is not inter-
ested in renewables. The United States in 2002 generated 2% of its
electricity from renewable sources. Most of this (71%) was from
biomass, 13% from wind and 16% from geothermal; less than 1%
was from photovoltaic cells.

The Public Utility Regulatory Policies Act of 1978 requires utilities to purchase power from certain qualifying non-utility producers, especially small (below 80 MW) renewables-based electricity
production at avoided cost rates from "qualifying facilities". The
interpretation of what constitutes an "avoided cost" was left up to
individual states. The qualifying facilities were also exempt from
some of the state and federal regulations that apply to utility generators. This legislative encouragement of non-utility generation has
contributed to increased electricity production from geothermal,
biomass, waste, solar, and wind. Biomass and waste-to-energy qual-
ify as long as they meet a 5% of useful steam threshold.

California interpreted "avoided costs" by forecasting higher
future energy prices than actually occurred; this created a favourable
investment climate for renewables in that state. In 1995, the Federal
Energy Regulatory Commission was given the responsibility for
defining avoided costs. It linked the costs to the costs that a utility
would incur either generating electricity directly or buying it from
another utility. This resulted in less favourable investment conditions
for the renewable industry.

In 1992, the Energy Policy Act (which has been renewed and extended over the years) provided electricity production tax credits for privately owned renewable generators. This has been particularly useful in encouraging wind turbines in the United States, especially at a time when wind turbines in the United Kingdom were virtually unknown. This incentive provided a wind turbine generating capacity of nearly GW in 2003, which was then double the capacity of wind turbines in Denmark.

Different states have a different approach to renewables, so it is particularly important to highlight where the Americans have made a renewable impact. In 2003, California, for example, generated 9% of its electricity from renewable sources – mainly geothermal and biomass. The state has always offered generous tax credits for renewables, funded turbine research, and sought to arrange a market where there was no price differential between renewable or fossil fuel generated electricity. In 1996, California provided useful net metering rules for domestic consumers who installed small scale photovoltaic or wind systems – something that we have yet to do in the United Kingdom.

> In 1998, California established a successful programme to help consumers and small businesses pay for the initial investment in renewables. In 1999, California provided a consumer credit for purchasing renewable electricity, currently at 1 cent per kWh with a limit of $1,000. This incentive is similar to the Dutch ecotax system, but different in that in Holland green electricity producers receive a production subsidy and Dutch green electricity is exempt from tax. This provides an incentive for consumers that "pulls" people into renewable electricity whereas the Californian system "pushes" people there, or seeks to.

The United States comprises 50 separate legal jurisdictions. That means that when we talk about "America" we must consider the many significant states programmes for renewable energy. There are incentives available in some states that go far beyond anything available to people in the United Kingdom and Europe. There are also very generous federal incentives in the form of tax credits.

Much of every nation's energy policy consists of what is, in effect, an electricity policy. The reason for this is that electrical generation is easily measured and is more polluting that heat generation. Heat generation is harder to measure and meter, so targets become more difficult to set. Nonetheless, probably around 40% of the energy used in the United States and Europe is used to create heat or cold, rather than power lights and appliances.

Most of the state incentive programs in the USA makes United Kingdom incentive programs look poorly thought-out and to be of poor incentive value. In the United Kingdom, the government has sought to use much more modest incentives or subsidies as a "pump primer". In the United States, the incentives are used to help over-come the difficulty of persuading people to invest in their own infra-structural energy generation in a way that they are not being overly disadvantaged for the common good.

In the United States, many of the incentives are in the form of tax credits. It seems to be accepted that tax incentives should be used to encourage activities that are for the common good. Most states in the US provide a property tax exemption, which means that if a householder improves his or her dwelling by a renewable installa-tion, the value of the improvement is disregarded. Many states exempt renewable energy creation systems from sales tax. No Euro-pean Union country exempts this equipment from value added tax.

In addition many states have a worthwhile and stable system of grants which can be accessed by individuals or not-for-profit organ-isations. These grants are not discretionary, as they have been in the case of not-for-profit organisations in the United Kingdom, but they are there as of right.

The federal government operates a system of energy credits, to be applied against energy bills, to encourage utilities to conserve energy. This compares favourably with the Energy Efficiency Com-mitment which, as we have seen, does not really produce, in my view, value for money in the United Kingdom. It also appears to have the same effect as is intended by carbon trading without the need to create an artificial market. The federal government also pro-vides tax credits for renewables, energy efficient mortgages, personal and corporate tax exemptions and other measures.

Individual states also have additional programs to encourage the use of renewables. Oregon, well known for its environmental poli-cies, has over 20 types of programs and rebates for many types of renewables. Businesses and consumers in Oregon are given plenty of help to move onto the renewable path.

Minnesota has recently commenced a program to use its arable crops to provide energy. It has done this by blending its petrol (it uses over 2.3 billion gallons of it each year) with 10% ethanol that is produced from crops. There are 13 ethanol plants in the state producing over 325 million US gallons of ethanol each year all from agricultural sources. The program will almost certainly grow, providing green energy as well as employment to thousand of people.

Texas, a great oil industry state, allows companies to deduct the cost of solar energy devices from taxable capital, or 10% from the company's income. It also exempts manufacturers of photovoltaic systems from tax.

Listing all the incentives available across the United States would be outside the scope of this work, but it is clear from the small samples I have examined that it cannot be seriously argued that the United States is against renewables, whatever the public perception may be.

On a federal level, we must also take into account the very important work that America has done in the field of energy efficiency of domestic appliances. The US system has involved expenditure by the federal government's Department of Energy of over 200 million dollars over the past 20 years. That money has been very well spent; consumers have on average saved $75 for every $1 spent by the program. 20 years ago, the federal government looked at domestic appliances, consulted with their manufacturers and set future minimum energy consumption levels for a range of appliances. These minima were legally binding; it is illegal to sell an appliance that does not meet its legal standard in energy efficiency.

These legally binding standards restricting power requirements of appliances are geared to come into effect within a stated period of years, which was enough for present products to finish their useful lives and for the manufacturers to redesign their products. The authority administering these standards also increased the range of appliances over the years and gave those that exceeded the efficiency minima the prestigious "Energy Star" award. This process continued over twenty years.

> Today in the US if you buy a 30 cubic foot refrigerator, it will use less energy to power it than a 10 cubic foot refrigerator needed in 1985.

It is clear that the United States may produce huge amounts of carbon, but this is to be expected from the largest industrialised country in the world. It is also clear from their opposition to the Kyoto Protocol that the US regards it as unfair that prosperous industrialised countries should suffer or risk large falls in living standards by implementing carbon savings when less prosperous nations should be allowed to pollute almost without control. I think there is force in that argument

Of course, as with every other country's efforts the judgement will be "must try harder"; it is however simply wrong to characterise the United States as not trying at all. The system of incentives seems to be one which we can all emulate. If Europe adopted the USA's minimum efficiency performance standards of appliances, huge carbon savings would be made and huge consumer energy expenditure would be saved.

Is the United States of America a villain or a hero when it comes to energy and global warming? In terms of the breadth of its measures and incentives it is heroic; in terms of actual emission creation it is a villain. But its villainy mainly happened when there was no knowledge of global warming. We cannot expect any nation to risk its prosperity when no other nation is prepared so to do, even for the sake of humanity and all our grand children.

Chapter 11

Energy use in the rest of the world

We have seen how energy is used in the United Kingdom, but of course, although the United Kingdom is a highly developed country, it is a small part of the world community. It consumes plenty of energy per capita, but relatively less than the United States, where the climate conditions are more extreme. We have reviewed some of the important features of energy use and policy in the European Union and in the United States of America. We will now look at how the rest of the world is using energy.

Let us first consider some of the facts; in 1970 the world, it is estimated, consumed 206 quadrillion British Thermal Units of energy. Two thirds of this was consumed by developed nations, with the United States being responsible for 30% of world energy consumption. Today world consumption is nearly double - probably around four hundred quadrillion.

Most predictions provide for the energy consumption growth of developed nations compounding at around 1% a year but for developing nations consumption compounding at over 5% a year. If you do the mathematics you find that by 2015 world consumption will have risen to over five hundred and sixty quadrillion. My own feeling is that a 5% compound rise is probably on the low side. Economic growth is a hungry beast that devours energy. A compound growth of 8% is not unlikely.

Wherever energy is used, consumption is on the increase and will continue to increase as undeveloped nations develop, and as developing nations reach developed status. It will also increase in developed nations as our apparently limitless thirst for consumption continues.

According to the Energy Information Association the increase in energy consumption in the twenty-year span from 1995 to 2015 will be equivalent to the total world energy consumption in 1970 – just before the oil crisis. Two thirds of this growth will, they predict,

Their only hope for a better life is that their countries will develop industries, services and productivity that will provide better employment and safer living conditions together with stability and peace. In order to do this their countries must use facilities, financial expertise and advice from the developed world. At the same time, the developing countries must provide energy for their own people so that their businesses and industries can flourish and the people live decently and healthily.

In these circumstances it would not be surprising to find the developing countries adopting indiscriminate energy policies so that they can grow more quickly. Although this has been the case in the past many states in the developing world are re-thinking their energy policy. As a result government intervention in energy markets is becoming increasingly influenced by concerns about global warming and the greenhouse gas implications of continued reliance on fossil fuels. Improved policies towards local air pollution are also causing change.

The energy consumption of many developing nations is concentrated in a few major industries that are highly energy intensive and often use energy inefficiently. This results from using less proficient technology, having smaller factories and failing to maintain energy-efficient operations within them. These industries usually produce chemicals, primary metals, cement, pulp and paper. They are all industries that are necessary for the development of infrastructure upon which economic growth is founded according to conventional economics.

China is the world's largest consumer of coal. As a consequence, China's particulate and sulphur oxide pollution is among the most extreme in the world. Until recently Chinese policy has encouraged the use of its coal resources. Pollution control technologies used in the Western industrialised countries to reduce the environmental impact of coal use are not widely used in China.

China has primarily used cleaner burning coal that is available in the northern regions of the country and is transported by rail to the southeast regions where the bulk of the economic development is taking place. However, the rail system carrying the coal is running at full capacity so the coal is being stockpiled at the mines. Abundant reserves of coal are available much closer to the south eastern economic growth areas, but the product here is brown coal with lower energy content and a much higher level of pollutants. What should China do? Should it expand its capacity to transport the cleaner fuel from the north, build electricity generation capacity in the north and a transmission system to bring the electricity to the south, or should it use the more polluting coal that is closer to the places where it will be used?

Notwithstanding this heavy use of coal, China has a highly developed thermal solar industry. Most of the solar collectors produced serve the home market, with only a fraction available for export.

China is now the second largest producer of greenhouse gases, after the United States of America. China produces about one seventh of the world's greenhouse gas each year. It is not, as we have seen, bound to any greenhouse gas reduction by Kyoto because it is a "developing" country. At the time of Kyoto, China indicated that it would not seek to reduce emissions until it had reached a stage of medium development, then expressed as a per capita income of $5,000 a year. It also expressed the view that developing countries must be allowed to emit greenhouse gases in order to improve their prosperity. This policy has a profound effect on world carbon dioxide emissions.

In the last few years China has changed from being a society based upon agriculture into an urbanised industrial power. In 1992 it exported coal; today it uses all its coal itself and imports more. In 2004 it had an energy demand of 1.9 billion tonnes of coal equivalent. This is expected to rise to 2.8 billion tce by 2015.

Despite refusing to ratify Kyoto, China is a leading manufacturer of renewable energy products which are mostly used in its own market. The reason for this is that China understands that dependency on imported energy is not the way to build a strong and safe economy. Renewable energy production is a way of diminishing albeit marginally, the need to import fossil fuel. For example, China has over 8 million square metres of solar collectors for water heating installed.

In February 2005 China enacted a Renewable Energy Law. This provides financial incentives to those developing wind, solar and bio-energy. The government allocated about $1.25 billion to subsidise the use of rural biogas. China's latest five year plan sets a target to save 240 million tce between 2006 and 2010.

It is important to understand what is happening in China because China has an energy intensity that is among the highest in the world. It is the most populated country in the world and it will inevitably be an important factor in what happens to the planet.

An interesting example of a developing nation that has suffered heavily from pollution caused by poor energy and environmental management is Mexico. In some parts of Mexico City – the world's largest city – the sewers are visible ten metres above the ground. It is one of the modern wonders of the world.

The sewers were laid below ground level in the1930s but Mexico City is sinking. So much water has been pumped out from the aquifer under the sewers to meet the needs of Mexico City's 18 million inhabitants that the land is sinking fast leaving behind the sewerage system firmly anchored in a hard layer of subsoil.

But Mexico is responding with more vigour and commitment to the present situation than many other nations. It has appointed an Environment Minister who has the equivalent of cabinet rank. He must be present when key cabinet discussions of any importance take place. He is not just an environmental figurehead but is a person who represents the environmental position in government; the importance that Mexico places on the need to consider policies from an environmental viewpoint is shown by the rank the minister has been given.

It is a mistake to see environmental issues as somehow being "outside" the economy. This is an error common both to those who advocate giving environmental issues very low priority in the interests of economic development, and to those who see economic development as being necessarily bad for the environment. Mexico recognises that good environmental management requires decision-making processes which incorporate consideration of all those potentially affected by any proposed policy. This can only happen where all issues, particularly energy issues, are considered not only from an economic viewpoint but also an environmental viewpoint. In democracies such consideration can only arise if a senior member of the government is appointed specifically and given befitting authority.

At the opposite side of the globe an often ignored nation is the eighth largest emitter of carbon in the world – Brazil. It is the home to over 180 million and has the world's largest forests, which soak up carbon dioxide.

Most of Brazil's greenhouse gas emissions do not come from energy production (unlike China) but from forestry and poor land use practices. However, as far as energy use is concerned, Brazil has advanced beyond many developed countries in some respects. Since the 1970s Brazil has been the world's largest producer of ethanol and has legally required it to be added to petrol. This simple meas-

Deforestation along the Highway BR-364 near Porto Velho in the Amazon. The BR-364 funnelled migration from the deserts of the over populated northeast. The migrant farmers clear forest to grow food but the land cannot sustain agriculture.

ure reduces both greenhouse gas emissions and pollution in the urban areas, where more than 80% of Brazilians live.

It refuses, however, to limit its greenhouse gas emissions. In fact it says that it will not limit its greenhouse gas emissions for another 40 years.

In the Pacific Ocean the Philippines (population of over 70 million) is growing at over 2.3% per annum. The country has a wealth of biomass, hydro and solar resources, and a potential wind resource in excess of 76GW capacity.   It has almost no oil resources but nevertheless oil will be its dominant energy source, followed by coal. There is also natural gas potential.

The Philippines generates 2,047MW each year from geothermal sources, making it second only to the USA in this field. It will shortly bring its energy production from hydro to around 3,813MW. At the moment its renewable energy is only 72 million boe.

The geography of the Philippines, with thousands of islands and many rural communities, has led the government to target rural electrification by using wherever possible locally renewable forms of energy.

Power lines in Bangkok, Thailand.

Thailand also has abundant renewable energy resources. Its Small Power Purchase Agreement supports the development of power projects to supply provincial and national electricity authorities from renewable resources. These electricity authorities are obliged to acquire energy from renewable energy projects under an agreed Power Purchase Agreement arrangement at agreed tariff rates.

Thailand has limited indigenous oil and gas reserves, and a large portion of its population can not plug into a power supply network. It will probably develop biomass as its leading renewable energy source; the country has a large agro-industrial sector farming rice, sugar and forestry. The government of Thailand has maintains support for energy conservation and renewable energy through the Energy Conservation Fund.

Australia is the leading user of energy on a per capita basis. Overall energy use was 3,158 PJ in 2000-2001 and will probably exceed 4,850PJ by 2020 increasing greenhouse gas emissions by at least 30%. 78% of the electricity is produced by coal. Australia has not ratified Kyoto and claims that it will meet Kyoto targets by "internal actions" rather than by Kyoto mechanisms. This is a valid policy because the Kyoto mechanisms are not the only way to achieve a reduction in the rate of carbon dioxide emission.

Australia has abundant supplies of fossil fuel resources. It has over 5% of the world's black coal reserves and over 20% of the world's brown coal reserves. It also has significant supplies of natural gas and over 40% of the world's uranium. Under Australian law, an additional 9,500 GWh of electricity must be generated from renewable resources by 2010. It also has significant programs and funds available for renewable energy – they are larger and more comprehensive than the United Kingdom's.

India, like China, is a critical energy user and will become a critical emission polluter. With large reserves of coal, and some oil and gas, India is a very rapidly developing economy. The government of India expects electricity production, standing at 336 TWh in 1997, to increase to over 1,300 TWh in 2020. Around 360 metric tonnes of coal were used in 1999; by 2020 the figure will be over 1,345mt. Over the same period petroleum products and natural gas will also be used in a way that that increases their consumption fourfold.

If there is to be a four fold increase in fossil fuel energy use by India by 2020 this will be far greater than all the Kyoto signatories' targets in the aggregate, assuming that they are all met. India is aware of this and has implemented encouragement to energy saving and conversation and program of modernisation of 44 thermal power stations. It is estimated that agricultural pumping units consume 28% of the electricity India generates. India has initiated a program of modernising and replacing these pump-sets which has led to a saving of 22MW. It has also made energy audit for businesses using more than 500KV a year compulsory.

Equally importantly it is replacing incandescent light bulbs with slim fluorescent tubes and encouraging the use of low energy light bulbs.

India is developing some critically important ideas about energy. Dr. R.K. Pachauri, Director of the Tata Energy Research Institute in New Delhi, believes that national security also involves the concept of environmental security. There are 2.8 billion people who live on less than $2 a day, and their environmental conditions and personal health are inexorably related to the lack of economic prosperity. Dr Pachauri feels that to attain environmental security a state must minimise environmental damage and promote sustainable development. Developing countries also usually lack the infrastructure and institutions to respond to crises, thereby increasing the likelihood of violent reactions.

School children measure the difference in energy requirements between a traditional tungsten bulb and a low energy bulb. Right: *Desalination plant powered by thermal solar in Oman. The whole world needs fresh drinking water and more desalination plants will have to be built as land use changes affect water quality and quantity.*

Pachauri has identified five areas where poverty has either exacerbated or been exacerbated by natural resource stress. First, the need to provide food, shelter and clothing is increasing land degradation. Second, worsening pollution affects the quality of the air. Third, world climate change has led to a rise in both temperature and sea level that will badly affect Asian coastal regions. Fourth, both the quality and quantity of water are at risk due to land-use changes. Finally, deforestation causes stress as forestlands are taken over for settlements, agriculture, and industry.

All these five "Pachauri areas" are directly related to energy consumption. Building on land reduces its vegetation cover and its carbon recycling capacity. Soil is degraded by energy-based pollution, so is air quality. We all understand the link between fossil fuel and climate change; water is often polluted as a result of energy use as well as land use changes, and trees are frequently cropped and burnt for fuel without there being a proper re-afforestation programme.

Pachauri argues that poverty is more than a mere lack of income. Poverty, he holds, is people's lack of ability to retain control over their living conditions. Something simple, like drinking water, is also a necessity of life and without it people cannot control their living conditions. Thus, Pachauri argues, if a community lacks empowerment to live in a way that is sustainable, poverty results. There are of course other factors that affect this but the core of his argument is very compelling as it points out a cycle between environmental degradation and poverty.

The planned economies of the former Soviet bloc evolved highly subsidised energy and water resources, which ultimately contributed to their very poor environmental records. The industrialised areas of communist Czechoslovakia and East Germany were the worst polluters in Europe and had the worst air quality in Europe. It is not only the planned socialist economies that have caused problems but also nations which have allowed industry to consume energy without any controls. Finally individuals, especially those who live in the high consuming prosperous Western economies, also contribute to the problem.

Against this background it is unsurprising that nations have conflicting interests in environmental energy use. Some need to maintain the standards and quality that they have created; others need to develop. There is inequality between nations and this makes it hard for genuine negotiations to take place. Institutional structures can make it difficult to find agreement or compromise between international organisations: between the International Monetary Fund and United Nations, for example or between national sovereignty. There is a need for an international organisation that has real teeth to enforce environmental protection.

In 1993 the World Energy Efficiency Association was founded as a private, non-profit organization composed of developed and developing country institutions and individuals charged with increasing energy efficiency. Its mission is to assist developing countries to access information on energy efficiency, serve as a clearing house for information on energy efficiency programmes, technologies and

Acid rain destroyed these Czech trees.

measures, disseminate this information worldwide, and publicise international cooperation efforts in energy efficiency. The WEEA points out:

"Modern energy can transform peoples' lives for the better. It improves productivity, frees millions of women and children from the daily grind of water and fuel wood collection, and through the provision of artificial lighting can extend the working day, providing also the invaluable ability to invest more time in education, health, and the community. Energy opens a window to the world through radio, television, and the telephone."

It highlights the facts that no country has managed to develop significantly more than a subsistence economy, without ensuring at least minimum access to energy services for most of its people, and that those living in developing countries attach a high priority to energy services. On average, these people spend, it is claimed, nearly 12% of their income on energy.

Providing energy to a population, especially energy created by the combustion of fossil fuels and biomass, will have adverse environmental effects. In rich countries, people are mainly concerned with the global consequences of fuel combustion, because many of its local effects (such as the infamous London "pea-souper" fog) have been controlled at considerable expense by clean air legislation.

In developing countries, the local environmental problems associated with energy use remain matters of concern that are at least as pressing as they were in industrialized countries fifty or more years ago. The poor suffer most severely from this type of pollution, because they have no access to better alternatives, and are forced to rely upon the most inefficient and polluting sources of energy, the WEEA claims, and further argues: "*World population is expected to double by the middle of the 21st century, and economic development needs to continue, particularly in the South . . . this results in a three- to five- fold increase in world economic output by 2050 and a ten- to fifteen-fold increase by 2100. By 2100, per capita income in most of the currently developing countries will have reached and surpassed the levels of today's developed countries. Disparities are likely to persist, and despite rapid economic development, adequate energy services may not be available to everyone, even in 100 years. Nonetheless the distinction between ``developed" and ``developing" countries in today's sense will no longer be appropriate. Primary and final energy use will grow much less than the demand for energy services due to improvements in energy intensities. We expect a one and a half- to three-fold increase in primary energy requirements by 2050, and a two- to five-fold increase by 2100.*"

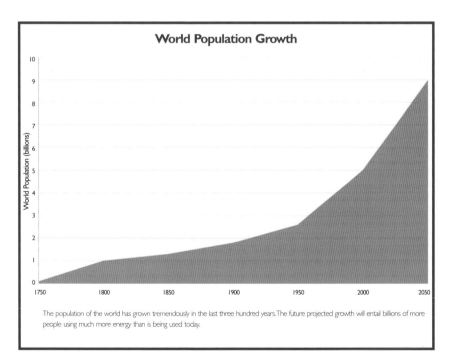

The population of the world has grown tremendously in the last three hundred years. The future projected growth will entail billions of more people using much more energy than is being used today.

The population of the world has grown tremendously in the last three hundred years. The future projected growth will entail billions of more people using much more energy than is being used today. What will happen to life on earth?

If the WEEA is right, the consequences for those alive in the next century will be drastic.

The means that the rich countries use to assist the development of poorer countries are loans and finance from the World Bank. This important institution attempts to provide effective development assistance to poorer countries and in doing so has an immense impact upon energy issues in the world.

Founded in 1944, the World Bank provided US$17.3 billion in loans to its "client" countries in 2001. It works with government agencies, nongovernmental organizations, and the private sector to formulate assistance strategies. More than 180 member countries own the World Bank and have the power to control its ultimate decision-making. Its main focus is on helping the poorest people and the poorest countries.

The World Bank's energy strategy is therefore vitally influential. The World Bank effectively decides which projects to fund and which to abandon. Its role is critical and its energy strategy of fundamental importance to the future.

There are three key parts to its strategy: policy assistance, knowledge management, and support for a variety of specific investments that help environmentally responsible policies, and support environmental best practice. The World Bank declares:

"The specific policy areas in which we will seek to engage clients and stakeholders are:

- Adopting a broad range of policies that target the principal sources of pollution across sectors, that are aimed at tangible improvements in environmental quality . . . and that balance the cost of compliance with environmental benefits.
- Accelerating the substitution of traditional fuels by modern energy and promoting new energy technologies, including renewables, by removing barriers to the development of their markets.
- Strengthening monitoring and enforcement capabilities for mitigating the environmental impacts of energy production and use across all levels of government, with a focus on local government and an increased role for communities and civil society.
- Promoting the restructuring of energy sector institutions and ownership as a focus of the energy-environment policy agenda, in order to capture the important environmental gains that energy sector reform involves. For example, pricing reform can enable proper reflection of the environmental costs of energy use and,

along with market liberalisation, can encourage improved energy efficiency.

The "clients" of the World Bank – the poorer nations that borrow - will inevitably consume more energy as they get wealthier. The World Bank's approach is different from Dr Pachauri's but both approaches agree on at least one thing – the need to promote energy efficiency and renewables is critical. For poorer people to become less poor and for the whole planet not to become impoverished it is of paramount and central importance that we use energy that is sustainable, renewable and non-polluting.

or the Future of Science". Russell's title sets the scene. If we, like Icarus, fly too close to the sun, we perish.

"Science", Russell said, "without altering men's passions or their general outlook, may increase their power of gratifying their desires". Russell was unsure whether, in the end, science will prove to have been a blessing or a curse to humankind.

In Russell's thinking science determined the importance of raw materials. Coal, iron and oil he regarded as the bases of power which in turn created wealth. He pointed out that countries that possess coal, iron and oil, could acquire markets by armed force, and economically dominate other nations. But coal and oil are only important for the energy they produce. So science enables us to increase our individual power of gratifying our desires. It does this by the use of energy; Russell asserted that the gratification science brought came without affecting the nature of man.

Human nature may or may not be changed by this process of endless self-gratification. To me this is not the most important point of the debate between the optimistic Haldane and the more critical Russell. It seems to me that the environment in which the human race lives will change. The change will occur because almost all the energy produced by humans is fossil based and creates by-products that will provide long term harm at the same time as they provide individual ephemeral gratification. Gratification inevitably presents the bill, which must be paid.

Bertrand Russell was not writing about global warming in his essay nor was Haldane. They both understood the importance, the overwhelming importance of energy. Russell was a pacifist; he drew an analogy, which we, in our present circumstances, would, however, appreciate in environmental terms: *"Some people think that we keep our rooms too hot for health, others that we keep them too cold. If this were a political question, one party would maintain that the best temperature is the absolute zero, the other that it is the melting point of iron. Those who maintained any intermediate position would be abused as timorous time-servers, concealed agents of the other side, men who ruined the enthusiasm of a sacred cause by tepid appeals to mere reason. Any man who had the courage to say that our rooms ought to be neither very hot nor very cold would be abused by both parties, and probably shot in No Man's Land. Possibly some day politics may become more rational, but so far there is not the faintest indication of a change in this direction."*

The Carmelite Monastery at Notting Hill London; the nuns grow much of their own food and try to live a sustainable life. The thermal solar panels on the slate roof provide them with hot water. (Future Heating Limited)

Russell concluded in his essay that science threatens to cause the destruction of our civilisation, and although he did not write from an environmental perspective, his words are beginning to ring true in the sense that science has enabled us to use energy as intensively and as greedily as we do. If we could adopt the vision of Haldane and produce our energy from renewable benign sources we can temper the pessimism of Russell.

One modern thinker, Brian Czech, has considered energy from its economic perspective. He points out that efficiency is not an absolute concept but a relative one and no where is this more important than in the field of energy. For example, Czech points out that if efficiency is defined as production or out put per person per hour, the American farmer is the most efficient, producing more per hour than any other farmer. However, if efficiency is defined in terms of production per unit of energy used (taking into account all the subsidies including energy subsidies that farmers receive) the American farmer is the world's least efficient producer, using much more energy per unit of production than any other.

Modern economics is obsessed with the marketplace. As a result in developed nations very few people are involved in actual production but their production feeds, clothes and shelters the majority of people who are involved in providing services. The use of energy makes this possible; no other factor does. There is no magic in the market place; if you remove cheap and freely available energy the nature of the market changes and fewer people will provide services and most will provide necessities.

Accordingly it has been suggested that the problem with acting in an environmentally friendly way is that to do so will reduce out put (as measured in terms of per person per hour) and that means that there is less production to support service industry. That means that instead of hiring services we shall have to provide some of them for ourselves some of the time if we are to develop a wholly environmentally friendly economic system.

Joseph Tainter has considered these issues from a historical view point. His studies of the collapse of ancient civilisations lead him to believe that they collapsed because their environments could no longer sustain their populations and this led to warfare, chaos conflict and ultimately collapse. He writes that all societies as they become more complex need to solve more complex problems which become more and more costly for them to attempt. They accordingly develop complex problem solving systems. *"In time such systems either require increasing energy subsidies or they collapse. Diminishing returns to complexity in problem solving limited the abilities of earlier societies to respond sustainably to challenges, and will shape contemporary responses to global change. To confront this dilemma we must understand both the role of energy in sustaining problem solving, and our historical position in systems of increasing complexity."*

There are, according to Tainter, two possibilities; over several generations energy shortages could lead to violence, starvation, chaos and ultimately a decline in population: or by using solar energy, renewable sources, and by consuming significantly less we might avoid a terrible future. However, if we are to avoid the apocalypse, we need to rethink whether economic growth is desirable and remove the concept of consumerism as beneficial to humanity.

We have all become or desire to become great consumers. All of our consumption requires energy, and yet so far, energy is fairly low on our personal agendas. Even today (and making an important exception for those in fuel poverty) most consumers in developed countries regard their energy bills, whether for the home or for transport, an inconvenience, and not an oppressive burden.

Electricity powers our computers, it enables us to play electronic games of a kind not dreamed of fifty years ago; it can provide the power to enable us to listen to music in near perfection and even to carry our music in our pockets. It has enabled us to access the internet, the world's cheapest information source. Electricity enables us to watch sporting or cultural events broadcast from the other side of the world. We can see our politicians argue debate and attempt to persuade us. We can watch wars as they unfold. Perhaps most disturbing of all, we can see massive tragedies, affecting one person or thousands of people, in all their horror.

We can have our illnesses cured, which were incurable a few years ago. We have eradicated some diseases and made others, once fatal, curable. We can extend life and extend quality of life, all by the use of science and energy.

We can drive to the other side of the country on good roads and in reliable cars. We can fly across oceans for less than an average person earns in four days. We can move food and goods from all over the world cheaply so that many countries develop a kind of manufacturing or service speciality in very product-specific fields.

Cheap gas enables us to live decently and warmly, to wash our clothes after we have worn them for very short periods. Only forty years ago in the United Kingdom, on Friday nights in most parts of the country, ordinary people went through the ritual of the Friday night bath; no family member was allowed to use up all the water and someone experienced in doing so would touch the bare copper of the hot water cylinder to try to establish whether the next person would have a sufficiently warm bath. Most people can afford to bath and shower when they want to because the cost factor is not significant to them.

Tainter fears that what will probably happen is that we will invest more and more resources in solving the more complex problems that our society will create. This will involve greater use of energy. *"This option is driven by the material comforts it provides, by vested interests, by lack of alternatives, and by our conviction that it is good. If the trajectory of problem solving that humanity has followed for much of the last 12,000 years should continue, it is the path that we are likely to take in the near future . . . This current default path leads to human extinction."*

We have two critical problems to solve involving energy. The first is one of climate change. The other is one of shortages. M King Hubbert was a geophysicist who created a mathematical model of petroleum extraction for any given oil field. This model indicated that over a period of time the total amount of oil extracted follows a logistic curve. The curve was a bell shaped one; after hitting peak the field would be depleted. He applied that modeling to the world supply of oil.

Hubbert predicted in 1956, that oil production would peak in the United States in around 1967 and that world production would peak in 2000. He was not far wrong, as far as US oil production, which peaked in 1971 but global production has not yet peaked, it would seem, although an increasing number of theorists believe that world oil product has already peaked. Similarly natural gas production has peaked in the United States and many developed countries. Of the major oil producing countries production has not yet peaked in Saudi Arabia, Iraq and Iran. Out of the world's leading 48 oil producing nations, 33 are experiencing a decline in oil production.

Oil, natural gas, uranium and coal are all natural resources available in finite amounts. At some stage their production will peak. If we have no alternative source of cheap energy we shall see nations competing for energy sources and such competition can lead to existing prosperous nations becoming impoverished and some nations, may seek to acquire energy sources by force, to protect the way of life of their inhabitants.

An increasing number of theorists believe some peak in world oil production has already occurred. The Association for the Study of Peak Oil & Gas has calculated that the global production of conventional oil peaked in the spring of 2004 albeit at a rate of 23-GB a year, not Hubbert's 13-GB a year. The Energy Information Administration and the International Energy Agency both predict no world peak in oil before 2025.

The Association for the Study of Peak Oil and Gas has proposed a protocol (known as the Rimini protocol) under which oil prices would be kept low and the peak effect deferred. Under this oil producing countries would have to agree not to produce more oil than their present rate of oil depletion and importers would not import more than their present rate of import. No nation has yet signed up to this idea. In effect it means rationing oil and finding the energy that oil has hitherto supplied from other sources.

In many ways the traditional sources of energy will become, as time goes by, undesirable and unfeasible or simply unavailable. There is much doubt about how much oil remains to be extracted from the earth. Supplies seem to be plentiful but things are not always as they appear. Do we have 50 years worth of supplies, 100 years worth or will increased consumption in Asia mean that we have less than 30 years of oil?

Similarly, coal resources must be finite. But regardless of the extent of all fossil fuel supplies, until the energy can be extracted from it without releasing carbon dioxide and other pollutants, burning fossil fuel will always be undesirable because it is environmentally harmful. The significance of environmentally harmful activity is not that it destroys some environments but that it ultimately destroys all environments and thus leaves us no place to live.

So, even if we do not believe that global warming will end our civilisation, or that economic growth will cause our civilisation to collapse inwards on itself because the product of the growth cannot be sustained, at some time in the not too distant future we shall run out of our traditional energy sources and that will cause our civilisation some difficulty.

When we start our working lives we are urged to plan for our retirement. That may be forty or so years on but it is still considered prudent to plan for the future and provide for it. As nations we spend so very little in planning our energy futures. There are some alternatives that we can consider.

The newest large source of energy is that created by nuclear power. In its creation the processes, provided they are rigorously controlled, are benign. The by-product is the problem. It is not possible to know that in, say, five hundred years we shall be able to contain the by-product successfully. Five hundred years is a very long time in history and as civilisations have regressed and forgotten technologies and practices of the past, we cannot be confident that they will retain the skills and desires needed to control the nuclear fuel by-product five hundred years in the future.

We can turn to biomass and similar forms of renewable energy. I think that the case for biomass is unconvincing. Burning any combustible substance creates carbon dioxide and the fact that the biomass may be renewed does not mean that we are placed in a better position. It seems that it would be far better to leave vegetation in place so that it can absorb carbon dioxide through the photosynthesis of plants and only burn it when it is no longer able to do this.

Wind energy is a form that many people hope will provide some solution to the problem of providing clean energy. A landscape covered with wind farms appears unusual and to some people unpleasant. We think that a prospect of wind farms stretching as far as the eye can see is no less intrusive than a landscape covered with traditional power stations together with their massive cooling towers. The windmills are sometimes noisier than we would like them to be and sometimes spoil pleasant views. These are important but not critical considerations as wind farms generate their power in a largely beneficial manner.

There are three more important considerations that apply to wind power. First, when they are combined in wind farms, windmills generate their power away from the places where the energy is to be used. This inevitably means that there will be a massive loss of power in transmission. Secondly, and perhaps more significantly, because a wind generator has many moving parts it is necessary to maintain them continually and this can be expensive, both financially and in the consumption of energy.

We have seen that ocean thermal power and tidal sources are unlikely to provide a solution to our problems and that heat pumps, are useful and effective provided that the electricity to power them is not too expensive.

Photovoltaic power (PV) may be considered as part of the overall energy production. PV installations are very expensive and inefficient when measured by Czech's definition. Between ten and twenty square meters of PV installation will provide an average home with between 30 and 65% of its electricity.

Even if vast economies of scale were possible and the price of PV could be reduced by 80% there would be some major disadvantages in PV. PV produces direct current electricity. We use alternating current and the DC electricity needs to be converted into AC electricity by means of an expensive inverter. In the process there are current conversion losses of between 7 to 12%. PV is by its very nature somewhat fragile. It is unlikely to be robust and results of accelerated aging tests are inconclusive. It is also fairly easily prevented from generating power if a small part of the installation becomes covered. PV probably recovers the energy used in its manufacture after around seven years and although the manufacturers claim a life expectancy of 20 years these claims have yet to be demonstrated.

We must also consider thermal solar technology. Even though some installations "merely" heat water, water heating is critical. In the United Kingdom the shortness of daylight in winter means that there is unlikely to be enough light to generate 100% of a family's hot water needs, but the 60%-80% that a good solar water heating installation can still give makes this form of energy viable and sustainable. If an energy technology driven by light and leaking very modest parasitic losses, like solar thermal, can be used it should be used everywhere. The fossil fuel savings are important.

So what then do we do in the future to secure our energy? We have to plan to use all the renewable technologies available so that they work together. We have to develop techniques to store energy better. We have to stockpile fossil fuel and secure good supplies of it. We have to conserve and use less energy. When we use fossil fuel we must try to sequestrate the carbon from the fuel as we use it. We have to use taxation as a way of financing rewarding and penalising energy use, according to its nature. These are the strategies available on national and local level.

As nations we have to stop looking for a single solution. We have to apply all the solutions and apply them all at once. And we have to stop pussy footing around.

First; the renewable technologies: let us build on the good research that has gone on and require all homes in developed countries to have at least two forms of renewable energy generation immediately.

We can start with the homes in the course of construction – systems can be easily and economically fitted to them. We do not allow these homes to be built without damp proof courses and safe electrical and gas services. Why do we allow them to be built without taking advantage of their capacity and potential to generate their own energy? We can shortly afterwards move on to the more expensive business of retrofitting these technologies. We have to accept that achieving a carbon neutral or an energy self sufficient home is going to be unlikely, but that should not be the object. Let us aim to produce homes that use from outside sources no more than 30% of their energy, with 70% of it being internally generated.

Once we have covered homes we turn our attention to legislating for factories and offices. This might be harder but we can build upon what we will learn when we tackle domestic energy users, which in most developed nations amounts to around a third of the carbon output and energy demand.

tudes to apartheid. Individuals are beginning to make their own small protests; some refuse to buy over packaged apples from the supermarket. Others refuse to buy drinks that are not packed in reusable glass bottles. In time, many people will refuse to buy goods made in third world countries where the manufacture of those cheap goods is done at the expense of the local and planetary environment. This may well prevent or restrict economic growth in those countries as we understand economic growth today.

In these days of free trade individuals can still chose not to consume; if consumption of a particular article is necessary and if consumers are choosing only those using energy efficiently and benignly the market will follow soon enough. Businesses only create carbon and pollute because their customers do not care. If the customers cared the businesses would stop polluting and created carbon.

I would encourage actions by individuals that boycott businesses and products that do not have a satisfactory energy profile. Human behaviour is at the critical factor in energy security and in climate change. The fact that developed nations have poisoned the planet for years is no reason to allow the under developed nations to poison it further.

I have no wish to present a "doom and gloom" scenario so often associated with the environmental agenda. The truth is that we do not know when the delicate balance of our atmosphere will be irretrievably poisoned by global warming any more than we know when fossil fuel will run out. However it is only prudent to look at the risks and attempt to manage them and reduce them as much as we can.

Is global warming any less a disaster if the effects are felt in one hundred years rather than in twenty? Is running out of energy sources less of a problem if we run out in two hundred years when the planet's human population may have increased tenfold or twenty fold? Why are we, as a race, so prepared to run unnecessary risks with our children's futures and their children's futures?

It is not foolish to manage risks and sacrifice a little of our self gratification; ultimately our way of life and life itself depends now and in the future on energy. We have to recognise this and adapt our lives, our economies and our politics to this inescapable fact lest, at some time soon, darkness falls, lest darkness falls.

Energy and power and how they are measured

The most important thing to understand is the difference between *energy* and *power*.

Energy is the ability to do work; energy is usually measured in units called joules (abbreviated to J) or it may be measured in units known as watt hours (abbreviated to Wh). *Power* is the rate at which energy is supplied – energy per unit of time. Power is measured in watts. One watt is equal to one joule supplied per second.

These units, when larger numbers are involved use the metric prefixes so there are kilojoules (KJ) (a thousand Joules) or mega joules (J) (a million Joules), giga joules (GJ) (a thousand million Joules). Similarly, a kilowatt hour (kWh) is a thousand watt hours, and a mega watt hour (Wh) is a million watt hours. A gigawatt hour is a thousand million watts (GWh) (10 to the power of 9 or 10,000,000,000). A tera watt hour is 10 to the power of 12 or 10,000,000,000,000. When it comes to the largest energy measurements the exajoule (joule x 10^{18}) and the petajoule (joule x 10^{15}) are used, abbreviated as EJ and PJ respectively.

Tera, mega, giga and exa are always abbreviated in upper case to T M G and E respectively. If you use lower case when you mean mega, it will indicate milli (10 to the power of minus three) which is a lot smaller than mega!

Watt hours are a very convenient way of measuring electrical energy. One watt hour is equal to a constant one watt supply of power over one hour. For example, if a light bulb is rated at 100 watts in one hour it will use 100 Watts. Over six hours it will use 600 watts. A low energy light bulb rated at 12 watts will over an hour use 12 watts and in six hours it will use 72 watts. A single kilo watt hour (kWh) is equivalent to 3.6 mega joules (MJ).

Originally the most widely used measure for heat was the British thermal unit (Btu). One Btu is the amount of heat needed to raise the temperature of one pound of water one degree Fahrenheit. Today, the Btu is mainly used in the United States of America. Europe and most of the rest of the world mostly uses kWh.

Heat is also measured in calories; one Joule is 0.239 calories.

1GJ is 0.948 million Btu or 239 million calories or 278 kWh. 1 Btu is 1055 Joules or 1.055 kJ. You may also come across Quads, which are a measurement usually used in Btu's. A Quad is a quadrillion Btu, (1015 Btu).

Measurements of power (that is to say the rate at which energy is supplied) are usually made in watts; one watt is 1.0 joule per second which is the same as 3.413 Btu per hour. The kilowatt, of course is 1000 watts; one kilowatt is 3,413 Btu per hour.

You will also come across measurements of power expressed in horsepower. One kilowatt is about 1.341 horsepower. A horsepower is 550 foot pounds per second which is the same as 2545 Btu per hour or 745.7 watts per hour. A foot pound is the energy required to push with a force of one pound for one foot (1 joule = 1 Newton metre, 1 joule = .737562139 foot pounds, 1 foot pound = 1.355817967 joules).

The world's energy consumption in 2003 was 409 EJ, of which fossil fuels provided 90% as primary energy. Of this 60 EJ was in the form of electrical energy, with only 10 EJ provided by nuclear generation.

Energy conversion factors for large quantities

To From	TJ	GCal	Mtoe	MBtu	GW
TJ	-	238.8	2.388×10^{-5}	947.8	0.2778
GCal	4.1868×10^{-3}	-	10^{-7}	3.968	1.163×10^{-3}
Mtoe	4.1868×10^4	10^7	-	3.968×10^7	11630
MBtu	1.0551×10^{-3}	0.252	2.52×10^{-8}	-	2.931×10^{-4}
GWh	3.6	860	8.6×10^{-5}	3412	-

TJ — terajoules

Gcal — gigacalories

Mtoe — millions of tonnes of oil equivalent

MBtu — millions of British thermal units

Gwh — gigawatt hours

Energy content of common fuels

Different fuels have different systems of measurements and differing energy contents.

Agricultural residues are measured by weight. The energy content of agricultural residues is between 10-17 GJ for every tonne (4,300-7,300 Btu per lb). The large range difference is due to varying moisture content.

Charcoal is usually derived from air dried wood. A metric tonne of charcoal yields 30 GJ (= 12,800 Btu/lb). To make a metric tonne of charcoal will usually require between 6 to 12 tonnes of air dried wood, which itself will have an original energy content of between 90-180 GJ.

Coal comes in various grades and qualities. A metric tonne of coal has around 27-30 GJ if it is bituminous/anthracite coal. If it is lignite/sub-bituminous coal the energy content is around 15-19 GJ. 27 GJ is 11,500 Btu per pound weight and 15 GJ is 6,500 Btu per pound. Most coal burning power stations in the UK use bituminous coal.

Ethanol is measured by weight. A metric tonne of ethanol is equivalent to 7.94 petroleum barrels which is 1262 litres. The energy content of ethanol is 11,500 Btu per pound weight or 75,700 Btu per US gallon or 26.7 GJ per tonne or 21.1 MJ/ per litre. HHV ethanol is 84,000 Btu/gallon or 89 MJ per gallon or 23.4 MJ/litre. A single metric tonne of bio-diesel contains 37.8 GJ (33.3 to 35.7 MJ per litre).

Natural Gas is usually measured and metered by volume. One cubic metre of commercial quality natural gas yields around 38 MJ (10.6 kWh), which is 1027 Btu per cubic foot.

Oil is measured in barrels. A barrel of petroleum oil is 35 Imperial gallons, which is 42 US gallons or 159 litres. A barrel of oil equivalent (boe) provides about 6.1 GJ (5.8 million Btu), which is equivalent to 1,700 kWh. A metric tonne of oil is about 7.2 barrels and provides 42-45 GJ.

Petrol or petroleum is sold by volume. In Europe it is sold by the litre but in the USA by the American gallon. A US gallon contains about 115,000 Btu or 121 MJ which comes to 32 MJ per litre (LHV). The energy content of petroleum products per unit mass is fairly constant, but their density differs significantly – hence the energy content of a given volume of petroleum product varies according to the product's density. This applies to products such as petrol, diesel and kerosene.

Wood is sometimes sold by the cord. A Cord is a stack of wood comprising 128 cubic feet (3.62 m3); standard dimensions are 4 x 4 x 8 feet, including air space and bark. One cord contains approx. 1.2 U.S. tons (oven-dry) which is the same as 2400 pounds or 1089 kg. A metric tonne of wood is 1.4 cubic meters of solid wood, not stacked wood. The energy content of bone dry wood is between 18 and 22 GJ per tonne (which is 7,600-9,600 Btu/lb). The energy content of wood fuel (air dried, having a 20% moisture content is about 15 GJ per tonne (6,400 Btu/ per pound weight)

Biomass sometimes requires processing before use, such as chipping of tree material, drying and pelletising of crops or the digestion of food or farm waste to produce methane. It is organic material produced by the photosynthesis of light.

Biomass boiler: A device for burning biomass to provide space and water heating to a whole dwelling (or to a collection of end-users via a heat network) on a controlled time and temperature regime, and with continuous fuel supply.

Biomass gasification: In gasification processes, wood, charcoal and other biomass materials are gasified to produce so called ´producer gas´ which can be burnt providing heat which can be used in itself or to drive turbines to generate electricity.

Biomass stove: A device for burning biomass to provide direct radiant heat to a single room.

Blocking diode: A solid-state electrical device planed in circuit between the module and the battery when the voltage of the battery is higher than that of the module (i.e. at night).

Boe: Barrel of oil equivalent.

Butane: Either of two saturated hydrocarbons, or alkanes, with the chemical formula C_4H_{10}. Butanes occur in natural gas, petroleum, and refinery gases. They are stable at normal temperatures but burn readily when ignited in air or oxygen.

By-pass diode: a solid-state electrical device installed in parallel with modules of an array which allows current to by-pass a shaded or damaged module.

Calorie: 4.1840 Joules (kilocalorie/kcal = 4.184 kilo Joules/kJ); be careful because there are five different types of calorie, all having slightly different measurements; we have given the S I value here.

Calorific Value (CV): The heat available from a fuel when it is completely burnt, expressed as heat units per unit of weight or volume of the fuel. The gross or higher CV is the total heat available; the net or lower CV is the total heat less the latent heat of the water vapour from the combustion of hydrogen plus any water in the original fuel. The power industry tends to use net CV, the boiler industry gross CV.

Candela: Unit of luminous intensity (cd). The candela is the basic unit of luminous intensity. It is the intensity of a source of light of a specified frequency, which gives a specified amount of power in a given direction.

Carbon: An element that combines with oxygen to form carbon dioxide. Emissions of carbon dioxide are measured in terms of the weight of carbon emitted. To convert tonnes of carbon (tC) into tonnes of CO_2 (tCO_2), multiply by 3.67.

Carbon dioxide (CO₂): Carbon dioxide contributes over 60% of the global warming effect of greenhouse gases caused by human activity. The burning of fossil fuels releases CO_2 fixed by plants millions of years ago and thus increases its concentration in the atmosphere.

Carbon Index (CI): The CI is based on the total annual carbon dioxide emissions associated with space and water heating per square metre of floor area. It is expressed as a number between 0.0 and 10.0 rounded to one decimal place. The Carbon Index can be used to demonstrate compliance with the relevant standard under the Building Regulations that apply in the United Kingdom.

Carbon monoxide: Oxide of carbon produced by fuel combustion. CO represents incomplete combustion and can be burnt to CO_2, which is complete combustion.

Carbon Storage: Long-term storage of CO_2 in the ocean or underground in depleted oil and gas reservoirs, coal seams and saline aquifers. Also referred to as engineered carbon sequestration.

Carbon Trust: An independent company funded by government to assist the UK move to a low carbon economy by helping business and the public sector reduce emissions and capture the commercial opportunities of low carbon technologies.

Cell (battery): The smallest unit or section of a battery that can store electrical energy and is capable of providing a current to an external load.

Cell (photovoltaic): see solar cell.

Celsius: A scale of measurements of heat. Freezing point of water at sea level is 0 and the boiling point of water at sea level is 100.

Centrally-generated electricity: Electricity generated in power stations and supplied via the National Grid.

CFL: Compact Fluorescent Lamp.

Chlorofluorocarbons: Gases made up of carbon, fluorine and chlorine, frequently in the past used as refrigerants or foam blowing agents. In the atmosphere they are broken down by strong ultraviolet light releasing chlorine atoms which recoat with ozone molecules in the atmosphere.

Charcoal: Material resulting from charring of biomass, usually retaining some of the microscopic texture typical of plant tissues; it consists mainly of carbon with a disturbed graphitic structure, with lesser amounts of oxygen and hydrogen.

Charge controller: A device which protects the battery, load and array from voltage fluctuations, alerts the users to system problems and performs other management functions.

Clear Skies: A capital grant scheme for most renewable technologies, open to householders and community groups in England, Wales and Northern Ireland. Scottish householders and not-for-profit community organisations can apply for grants from the Scottish Community and Household Renewables Initiative.

Climate: climate is usually defined as the "average weather", which in turn means using statistics to describe weather (temperature, precipitation and wind) in terms of the mean and variability over a period of time. The World Meteorological Organization use 30 year periods but periods can be as short as months or as long as tens of thousands of years.

Climate change: Climate change refers to a statistically significant variation in either the mean state of the climate or in its variability, persisting for decades or longer. Climate change may be due to natural internal processes or external events, or to persistent anthropogenic changes in the composition of the atmosphere or in land use. Article 1 of the Framework Convention on Climate Change, defines "climate change" as: "a change of climate which is attributed directly or indirectly to human activity that alters the composition of the global atmosphere and which is in addition to natural climate variability observed over comparable time periods". This distinguishes human activity caused climate change from climate change caused by external events – such as meteors or volcanic explosions. Today most commentators use climate change in the sense defined by Article 1, and not its wider sense.

Climate Change Levy (CCL): A tax applied to the energy use in many non-domestic sectors, (but there are extensive exemptions) the purpose of which is to encourage energy efficiency.

Climate Change Programme (CCP): The UK's contribution to the global response to climate change. It sets out a strategic package of policies and measures across all sectors of the economy.

Cluster-LED: Recently-developed lighting technology based on output of Light Emitting Diodes. Cluster LEDs are more efficient and long-lasting than fluorescent lamps. The lighting is highly directional.

CO_2: Carbon dioxide most usually produced by complete combustion of fuel.

Combined Cycle Gas Turbine (CCGT): Combined cycle gas turbines use both gas and steam turbine cycles in a single plant to produce electricity with high conversion efficiencies and relatively low emissions.

Combined heat and power (CHP): Combined Heat and Power is the simultaneous generation of usable heat and power (usually electricity) in a single process. This means that there is less waste heat than conventional generation.

Community heating: A community heating system provides heat to more than one building or dwelling from a central plant via a heat network.

Compact fluorescent lamp (CFL): Mostly called low energy light bulbs because they use significantly less electricity to provide light than conventional tungsten bulbs.

Condensing mode: the efficiency of a boiler can be improved if it is designed for and operated in 'condensing mode'. In the right conditions, extra energy is retained in the heating system as water condenses (thereby giving up some heat), rather than all being lost in exhaust gases.

Connector strips: insulated screw-down wire clamps used to fasten wires together in PV systems.

Contraction and convergence: The panacea for climate change proposed by the Global Commons Institute. "Contraction" is the process under which all governments are collectively bound by an upper limit to greenhouse gas concentration in the atmosphere. "Convergence" is the process whereby developed and developing countries converge on the same allocation per inhabitant by an agreed date.

Converter: An electronic device for DC power that steps up voltage and steps down current proportionally (or vice-versa).

Compression-Ignition System: Used in reciprocating engines whereby fuel is injected after compression of the air and is ignited by the heat generated by compression. As pre-ignition is thereby eliminated, higher compression ratios than with spark-ignition engines can be utilised, with corresponding high energy conversion efficiency.

Crystalline silicon: A type of PV cell made from a single crystal or polycrystalline slices of silicon.

Current (amps amperes, A): Electricity current is the rate at which electricity flows through a circuit, to transfer energy. Current is measured in Amperes ("A"), commonly called Amps.

Cycle life: of a battery, the number of cycles it is expected to last before being reduced to 80% of its rated capacity.

Cycle: One discharge and charge period of a battery.

Cylindrical Rotor Generator: A type of electricity generator. As frequency depends on the speed multiplied by the number of pole pairs, higher speeds require fewer poles and the exciting winding can be accommodated in radial slots machined into the periphery of the rotor.

Daily energy requirement: The amount of energy that a household, requires to meet the sum total of its energy needs. The daily energy requirement for a typical European household is approximately 45 kWh but the typical energy requirements for an average East African rural household is on the order of less than 2 kWh.

Daytime valley: A reduction in the national electricity demand profile in between the morning and evening peaks.

DC – Direct Current: DC is the type of power produced by photovoltaic panels and by storage batteries. The current flows in one direction and polarity is fixed, defined as positive (+) and negative (-). Nominal system voltage may be anywhere from 12 to 180V. See voltage, nominal.

Decent Homes Standard: a standard or requirement set by the ODPM, the Decent Homes Standard is a minimum standard that all social housing in England should achieve by 2010. A decent home is 'wind and weather tight, warm, and has modern facilities'. The standard has been widely criticised as not being high enough.

Deep discharge battery: A type of battery that is not damaged when a large portion of its energy capacity is repeatedly removed (i.e. motive batteries).

DEFRA: UK Department of the Environment, Food and Rural Affairs.

Delivered energy: Energy supplied to a customer. Also referred to as 'energy supplied'. See also primary energy and useful energy.

Demand, Maximum Demand, Demand Profile: The rate at which electrical energy is required, expressed in kW or MW. It is usually related to a time period. Maximum demand is the highest half hourly rate at which electricity is required during a month or year. Peak load or peak demands are the terms usually used for heat energy. A graph of demand rate over a typical day, for example, is the demand profile.

Depth of discharge: A measure in percentage of the amount of energy removed from the battery during a cycle.

Desertification: Land degradation in arid, semi-arid, and dry sub-humid areas usually resulting from climatic variations and human activities. Soil erosion, soil deterioration and long term loss of vegetation usually cause this.

Diesel Engine: Compression-ignition reciprocating engines, usually but not always powered by diesel.

Diffuse radiation: Solar radiation that reaches the earth indirectly due to reflection and scattering.

Direct current (DC): Electric current flowing in one direction.

Direct radiation: Radiation coming in a beam from the sun which can be focused.

Distributed generation: Electricity generation plant that is connected directly to distribution networks rather than to the high voltage transmission systems.

District Heating Systems: A system which enables a large number of homes buildings and establishments to receive their heat and hot water from a single source, rather than by having individual heat making appliances in situ. The world's largest district heating system is at Marstal, on the island of Æro in Denmark and comprises a very large field of solar collectors supported by a back up biomass boiler.

Diurnal temperature range: The difference between the maximum and minimum temperature during a whole day.

Domestic Energy Efficiency Scheme (DEES): The Northern Ireland equivalent of Warm Front.

Domestic Tradable Quota (DTQ): See Personal Carbon Allowance.

DTI: UK Department of Trade and Industry.

Dual-fuel: the use of two fuels in a prime mover or boiler.

Ecosystem: A complex system of life forms living together (and off each other) in a community.

Efficiency: Efficiency is the percentage of power that gets converted to useful work. A system that is 60% efficient converts 60% of the input energy into work; the remaining 40% becomes waste. Efficiency when applied to renewable energy systems, such as thermal solar or PV, is not a very helpful concept because the source of the fuel – the light – is free. When applied to biomass boilers or fossil fuel systems it is an important concept. Efficiency can be measured in economic terms as unit output per hour or unit output per unit of energy.

El Niño: A warm water current which periodically flows along the coast of Ecuador and Peru. There is a fluctuation of the inter-tropical surface pressure pattern and circulation in the Indian and Pacific oceans. The prevailing winds weaken and the equatorial counter-current strengthens, causing warm surface waters in the Indonesian area to flow eastward to overlie the cold waters of the Peru current, causing great climatic changes in many places in the world.

Electricity: Energy that is generated by mechanical, thermal or photovoltaic means, and used to provide power for a number of applications.

Electric power: the rate at which energy is supplied from an electricity generating source. It is measured in watts (W).

Embedded generation: An alternative term for distributed generation.

Embedded Energy: This term is often used to describe the amount of energy used to create a product. Renewable energy products, such as thermal solar panels or PV panels and wind turbines, should be designed to recover their embedded energy quickly. If they cannot recover the embedded energy over the product life span they are pointless, in terms of providing renewable energy.

Emissions factor: The carbon emitted as a by-product of generating one kilowatt-hour of energy from a fuel or mix of fuels. Different electricity generators are brought on-line to meet peak demand, so the overall fuel mix typically changes, resulting in different emissions factors at different times of day and at different seasons. It is usually expressed as kilograms of carbon (or carbon dioxide) per unit of delivered energy (kgC/kWh or $kgCO_2$/kWh).

Energy: Energy is the product of power and time, usually measured in Watt-hours.

Energy Efficiency Commitment (EEC): The Energy Efficiency Commitment is a legal obligation placed on all domestic energy suppliers to achieve a specified energy saving target through the installation of energy efficiency measures in homes across Great Britain. It is funded by the customers of the energy suppliers, through their bills. Legally half of the benefits must be provided for vulnerable households. A similar scheme (the Energy Efficiency Levy) operates in Northern Ireland. The main energy saving measures are insulation (loft and cavity wall) and low energy bulbs.

Energy balance: Averaged over the globe and over longer time periods, the energy of the climate system must be in balance. A climate system derives all its energy from the sun. Therefore the amount of incoming solar radiation must on average be equal to the total of the outgoing reflected solar radiation plus the outgoing infrared radiation emitted by the climate system. A distortion of this balance is called radiative forcing, which can occur by natural or human induced activities.

Energy Conservation Authority (ECA): One of the 408 local authorities in Great Britain responsible for reporting on all the housing in their area under the Home Energy Conservation Act. The Housing Executive for Northern Ireland acts as ECA for the whole province.

Energy Efficiency Standards of Performance (EESoP): The precursor of Energy Efficiency Commitment.

Energy Intensity: the ration of energy consumption to GDP. This is normally expressed as Mtoe per US$billion. It is used primarily as a way of comparing energy usage in different countries.

Energy-using Products Directive: a European Union framework directive to reduce the environmental impacts of energy-using products (except vehicles), paving the way for the development of eco-design requirements on a product-specific basis.

Energy Performance of Buildings Directive (EPBD): This European Union directive requires each member state to: establish a methodology for rating the energy performance of buildings; energy certificates are to be issued when a building is built, sold or rented; an inspection regime is to be created for large energy installations in buildings; renewable technologies are to be considered when a new building is being designed.

Energy Saving Trust (EST): The Energy Saving Trust was set up by the UK Government following the 1992 Rio Earth Summit, with the object of achieving sustainable and efficient use of energy, and cutting carbon dioxide emissions from the residential sector. The EST is a non-profit organisation.

Energy Service Company (ESCo): An organisation that can provide energy supply (both conventional and low and zero carbon) and demand management to enable implementation of the least cost option, taking into account the cost of borrowing.

Energy from Waste: Electricity or heat, generated from municipal waste. Cleaner techniques than incineration include pyrolysis or gasification of waste.

Entropy: a measure of the amount of energy in a closed system that is no longer available to effect changes in that system. A system becomes closed when no energy is being added or removed and energy becomes unavailable by becoming disordered. In an open system matter tends to equalize their thermodynamic states thereby reducing differentials toward zero. Pressure differences, density differences, and temperature differences, all tend toward equalizing. Entropy is a measure of how far along this process of equalization has come. Entropy increases as this equalization process advances.

Exajoule: (EJ) a joule x 10^{18}.

Excess Air: Reciprocating engines and gas turbines have to operate with far more air than is needed purely for the combustion of the fuel. This excess over requirements forms the major proportion of the exhaust gases and is termed excess air.

Farad: A unit of the capacitance of an electrical system, that is, its capacity to store electricity.

Fahrenheit: A scale of measurements of heat. The freezing point of water at sea level is 32 and the boiling point of water at sea level is 212.

Fluorescents: light emitted from special inert gases (generally neon) when an electric current is passed through it. Fluorescent lamps are much more efficient than incandescent lamps, and are preferred over incandescent lamps for energy efficiency. They are not as efficient as halogen or LED lighting.

Fischer-Tropsch process: A way of making synthetic petroleum by converting carbon dioxide and monoxide and hydrogen into liquid hydrocarbons using iron or cobalt as catalysts. The carbon dioxide and monoxide is generated by using coal or wood. One of the largest plants existed in Most, in the Czech Republic and was in production throughout the 1930s and 1940s. The process is today used in South Africa to produce diesel. Over the whole life cycle fuel from the process produces more than double the carbon emissions of conventional fuel.

Force: The SI unit of force is the *newton.*

Frequency: The number of times per second that alternating current changes direction. Frequency is expressed as cycles per second or Hertz (Hz) of alternating current. The electricity supply in Europe is at a frequency of 50 Hz. In North America the frequency is 60Hz.

Fuel Poverty: A household is in fuel poverty if, in order to maintain a satisfactory heating regime, it has to spend more than 10% of its income (including Housing Benefit or Income Support for Mortgage Interest) on all household fuel use. See National Energy Action.

Fuel cell: A fuel cell produces electricity in a chemical reaction combining hydrogen fuel and oxygen (present in the air). Hydrogen is often extracted from natural gas (CH_4). There are many different designs of fuel cell, each with advantages and disadvantages.

Fuse: a device which protects circuits and appliances in the system from damage by short circuits.

g: gram (a unit of weight).

Gasoline: (petrol) A mixture of the lighter liquid hydrocarbons, used chiefly as a fuel for internal-combustion engines.

Generator: An alternator or direct current (DC) generator. "Generator set" refers to the combination of prime mover and generator.

Genset: a piece of equipment that generates electricity. Genset is short for generating set (*gennys* in North America). Often diesel powered generators are used for back-up power for hospitals, hotels, industries, but also used as primary electricity sources for isolated, off-grid applications, and in hybrid applications.

Geothermal energy: Energy that is generated by the heat of the earth's own internal temperature.

Giga: one thousand million units (in US terms, one billion units; in most European languages, giga = milliard).

Greenhouse gas: A greenhouse gas is one that contributes to global warming. The most significant are carbon dioxide, methane and nitrous oxide. Water vapour is also a green house gas. The Kyoto Protocol identified these as Carbon dioxide (CO_2) Methane (CH_4) Nitrous oxide (N_2O) Hydrofluorocarbons (HFCs) Perfluorocarbons (PFCs) and Sulphur hexafluoride (SF_6).

Greenpeace: a non profit international organisation that focuses on what it regards as the most crucial worldwide threats to biodiversity and environment; at the top of their list of threats is climate change.

Grid: A network that connects a supply of electricity to users. It consists of some form of electricity generator, with electricity taken along a transmission line at high voltage, and then stepped down to lower voltage on a distribution system that delivers electricity to end users.

Gross Domestic Product: the total market value of goods and services produced in a given period of time.

Halogen lamps: Lamps with very low wattage, that generate high intensity light through a combination of specially coated, highly efficient reflectors.

Hand pump: A pump driven by human force, generally by hand, or by foot, or sometimes, by the weight of the human body.

Heat Exchanger: A device in which heat is transferred from one fluid stream to another without mixing. Heat exchange operations are most efficient when the temperature differentials are greater. Heat exchange has the advantage of keeping potable water separate from water passing through a boiler or a thermal solar system, thereby ensuring the quality and safety of the potable water. A coil in a domestic hot water cylinder is an example of a heat exchanger.

Heat Grade: A classification of heat source or heat requirement according to temperature. Up to 50°C is classed low grade; medium grade is between 50°C to 150°C and high grade heat is above 150°C.

Heat network: A system of pipes taking heat, typically in the form of hot water, from a centrally sited energy centre to any number of homes, or other end-users. Sometimes call a thermal ring main, used in District Heating Systems.

Heat Power Ratio: The amounts of heat energy and electricity produced by a Combined Heat and Power boiler or appliance, expressed as a ratio. There is no agreed standard.

Heat pump: A device to exploit heat differentials. Heat pumps work like refrigerators, moving heat from one place to another. Energy can come in the form of electricity (such as vapour compression) or be thermal energy (absorption heat pumps). Heat pumps can provide space heating, cooling, and water heating and exhaust air heat recovery.

Heat recovery: A technique for maximising efficiency by making use of heat that would otherwise be wasted.

Hertz: The hertz is the SI unit of the frequency of a periodic phenomenon. One hertz indicates that 1 cycle of the phenomenon occurs every second.

Home Energy Conservation Act (HECA): UK legislation that requires all Energy Conservation Authorities to report annually on the energy efficiency of the housing stock in their area.

Home Energy Efficiency Scheme (HEES): A grant scheme for low income, fuel-poor households to fund a range of insulation and heating measures. The name HEES is still in use in Wales, but has been superseded in England (Warm Front), Scotland (Warm Deal) and Northern Ireland (Warm Homes).

Housing Health and Safety Rating System (HHSRS): Separate hazards in a dwelling are weighted according to the harm that could result. Excessive cold and high temperatures are classified as potentially hazardous to the elderly but to no other age group. The 'ideal' is stated to be a SAP rating of 80-85, with minimum and maximum temperatures of 16.25°C and 25.25°C.

Hydraulic ram: A device that uses the energy of falling water to lift a lesser amount of water to a higher elevation. A hydraulic ram uses a small fall of water to lift a small part of the overall water supply to a much higher height. Mechanically, they require just two valves. As the flow of water increases through the system opens and closes one valve causing hydraulic pressure to force some of the water upwards. Modern hydraulic rams use an air chamber to make the device work more efficiently and to cushion the system from hydraulic shocks.

Hybrid system: A hybrid system refers usually to the combination of two energy generating applications (e.g., diesel generator with PV system) to provide electricity at all times, or in all critical times. Generally, a hybrid system will be designed to ensure reliability.

Hydropower: power generated by the flow of water.

Illuminance: Luminous flux per unit area. Unit is a lux which equals a lumen per m^2.

Illuminous efficacy: Efficiency with which a surface is lit. Unit is lux per Watt.

Incandescent: an incandescent lamp produces light when its wire filament is heated by electricity to 'incandescence. The wire filaments are made of tungsten. The traditional tungsten-vacuum light bulb is highly energy intensive.

In-duct Burner: A burner sited inside the duct of the air or gas stream it is heating, and thus also adding its combustion products to the stream.

Insolation: A measure of the solar energy incident on a given area over a specific period of time. Usually expressed in kilowatt-hours per square metre per day.

Interconnected system (ICS): an electrical grid.

Intergovernmental (The) Panel on Climate Change (IPCC): The IPCC has been established to assess scientific, technical and socio- economic information relevant for the understanding of climate change, its potential impacts and options for adaptation and mitigation. It is open to all Members of the UN and of WMO. It is a body that assesses scientific research, rather than carries out its own research and is currently chaired by Dr R Pachauri.

Intermittence: when an energy source, such as sunshine, wind, even hydro or diesel, is not always available. Any system needs to be designed on the basis of how much constant, steady energy is needed, and whether demand can be served with intermittent supplies.

International Energy Agency: (IEA) An autonomous international body established in 1974 to produce and collate key energy data and promote energy stability and conservation. Notwithstanding its autonomy it is part of the Organisation for Economic Co-operation and Development. All the major developed countries are members.

Inverter: An electronic device that converts low voltage DC to high voltage AC power. In solar-electric systems, an inverter may take the 12, 24, or 48 volts DC and convert it to 115 or 230 volts AC, conventional household power.

Irradiance: the solar radiation incident or a surface per unit time. Expressed in watts or kilowatts per square metre.

Joule: A joule is the amount of work done when an applied force of one newton moves through a distance of one metre. A foot pound is the energy required to push with a force of one pound for one foot (1 joule = 1 Newton metre, 1 joule = .737562139 foot pounds, 1 foot pound = 1.355817967 joules). The joule is the SI unit of work or energy.

Kelvin (K): a scale measuring temperature: Freezing point is 273.15 degrees K = 0 degrees Centigrade or otherwise expressed as the basic unit of temperature. It is 1/273.16th of the thermodynamic temperature of the triple point of water.

Kerosene: A light fraction petroleum product refined from the raw petroleum. Kerosene is one of the lighter "distillates" in a petroleum refinery, lighter than gas oil/diesel, and often in the same mix with jet fuel. It has been used for lighting, cooling and refrigeration for one hundred years. Kerosene is found throughout the world, and is one of the most common lighting fuels in the developing world. It is also often used for cooking and water heating primarily in urban areas in the developing world.

Kilocalorie (kcal): A measure of energy equivalent to 4.187 kJ/kilojoules, or 1.63 Wh/Watt hours.

Kilogram: One thousand grams (1000g). The kilogram is the basic unit of mass. It is the mass of an international prototype in the form of a platinum-iridium cylinder kept at Sevres in France.

Kilo: One thousand units. one kilowatts = 1,000 Watts; one kilojoule = one thousand Joules).

Kilovolt ampere (kVA) (one thousand Volt Amperes): The current flowing in a circuit multiplies by the voltage of that circuit, usually measured on a transformer.

Kilowatt-hour (kWh): A kilowatt hour is 3.6 MJ. It is also an energy measure that indicates a Watt consumed or generated in one hour equivalent.

Kilovolt (kV): 1000 volts.

Kilowatt, kilowatt electric: kW is used to express energy rate of production or demand, whatever the energy form entailed.

Kyoto Protocol: The Kyoto Protocol binds those industrialised nations that are signatories to reduce emissions of greenhouse gases by an average of 5.2% below 1990 levels by 2008-2012. The UK is legally bound by its Kyoto target to reduce greenhouse gas emissions by 8% over that period. The protocol came into force on 16th February 2005.

LDC: a term used for 'lesser-developed country'. This term has largely fallen out of use, and the term 'developing country' tends to be preferred.

Light-emitting diode (LED): a type of semi conductor diode which lights up when current is flowing through it. It is highly efficient and long-lived, and has a low power requirement. Likely to be a future low carbon lighting product.

Liquefied petroleum gas (LPG): A mixture of butane, propane and other light hydrocarbons derived from refining crude oil. At normal temperature it is a gas but it can be cooled or subjected to pressure to facilitate storage and transportation. Often used for water heating in hot sunny countries.

Load: The amount of demand placed on an energy system. In the case of most electricity, load could be the set of equipment appliances that use the electrical power from the generating source, battery or module, and the amount of electricity (the load) that those appliances require. Load is often used synonymously with "demand". Load is usually expressed in "watts", so that, for example, if a refrigerator has a rating of 1 kW, the load is cited as being a 1 kW load.

Load factor: Usually applied to generating plant, load factor is the ratio of the average electrical load to the theoretical maximum load, expressed as a percentage. The UK's overall power plant load factor has been consistently rising over the past 20 years making power cuts much more likely. "Load factor" is also used to describe the average intensity of usage of energy producing or consuming plant expressed as a percentage of its maximum rating.

Low Temperature Hot Water: Hot water at up to 100°C used for space heating and/or for low temperature processes.

Low or zero carbon (technologies: Low or zero carbon technologies are taken to be renewable energy generators or technologies with better fuel efficiency than conventional technologies, and which are retrofitted to or integral to the building or community. Over these, solar water heating, PV and wind are virtually carbon free. These technologies are often described as LZC.

Mega: equals 1 million units (106). Example: one megawatt (MW) = one million Watts (106 Watts); one megajoule (MJ) = one million Joules (106 Joules).

Methane: A gas composed of carbon and hydrogen, the first member of the paraffin or alkane series of hydrocarbons. It is lighter than air, colourless, odourless, and flammable. Methane occurs in natural gas, in coal mines, and as a product of decomposition of matter.

Metre: The metre is the basic unit of length. It is the distance light travels, in a vacuum, in 1/299792458th of a second. This is roughly three feet three inches and four tenths of an inch.

MJ: Mega Joule = one million Joules (106 Joules).

Medium Temperature Hot Water: Pressurised hot water at 100°C to 150°C used for space heating or for processes.

MJ, GJ: Mega joules, Gigajoules. Units expressing quantity of heat energy. One GJ is roughly equivalent to 10 therms or 280 kWh.

MtC: a million tonnes of carbon being carbon dioxide weighed as carbon only.

MW, MWe, Megawatt, megawatt electric: As for kW, kWe, but 1 MW = 1000 kW.

Municipal Solid Waste (MSW): All rubbish collected by local authorities (or their contractors), including rubbish from homes, schools, colleges and co-collected trade waste.

National Energy Action: (NEA) a charity that seeks to highlight the plight of those in fuel poverty and help the UK government deliver its statutory duty to abolish fuel poverty.

Network: The distribution system which links energy production to energy usage.

Newton (N): The newton is the SI unit of force. One newton is the force required to give a mass of 1 kilogram an acceleration of 1 metre per second per second.

NGO: Non governmental organisations.

NOx: Nitrous oxides. A general term for oxides of nitrogen produced by fuel combustion, eventually discharged to atmosphere and considered deleterious emissions.

OECD (Organisation for Economic Co-operation and Development): comprises Australia, Austria, Belgium, Canada, Czech Republic, Denmark, Finland, France, Germany, Greece, Hungary, Iceland, Ireland, Italy, Japan, South Korea, Luxembourg, Mexico, Netherlands, New Zealand, Norway, Poland, Portugal, Slovakia, Spain, Sweden, Switzerland, Turkey, United Kingdom, United States of America.

ODPM: UK Office of the Deputy Prime Minister, responsible for policy on housing, planning, devolution, regional and local government and the fire service.

Office of Gas and Electricity Markets (OFGEM): is the UK energy regulator. Ofgem oversees the operation of the Energy Efficiency Commitment. Ofgem provides virtually no encouragement or support for renewables.

Off-grid: A situation when a consumer is not connected to an electricity grid. This is fairly rare in Europe or North America (limited to camping areas, very isolated sites,), but is very common in rural areas, and many near-urban (peri-urban) areas of the developing world where electricity companies and suppliers have been unable to connect domestic, commercial, industrial and institutional consumers. The main thrust of this workbook is to provide "off-grid solutions".

Ohm (SI): a unit of electrical resistance.

Open circuit voltage (Voc): the maximum possible voltage across a solar module or array.

Parasitic: Adjective for the energy used within the plant or equipment (such as a solar system) itself for the process and therefore reducing the amount available for beneficial use or being needed to balance against the amount of energy produced, if the source energy is different from the produced energy.

Particulates: Particles of solid matter, usually of very small size, derived from the fuel either directly or as a result of incomplete combustion and considered deleterious emissions.

Pascal (Pa): The Pascal is the SI unit of pressure. One Pascal is the pressure generated by a force of 1 Newton acting on an area of 1 square metre. It is a rather small unit as defined and is more often used as a kilopascal [kPa].

Pass-out steam: Also called extraction steam. Steam taken part-way along a steam turbine to serve a requirement for that particular pressure, the balance remaining in the turbine to the exhaust stage to generate more power. There may be more than one pass-out tapping to serve differing site requirements.

Payback: The capital cost of a device can be related to the energy savings it makes in terms of a payback period. The overall energy savings should be brought into account and any indirect savings as well as the cost of the finance and the overall product life. Fossil fuelled devices offer no pay back at all.

Peak power (Wp): the amount of power a solar cell module can be expected to deliver at noon on a sunny day (i.e. at Standard Test Conditions) when it is facing directly towards the sun.

Personal Carbon Allowance (PCA): Under a scheme that has been proposed of PCAs, each person would be allowed to use an equal amount of carbon emissions generated from fossil fuel. Allowances would be tradable, and would decrease over time. The aim of the scheme would be to deliver guaranteed levels of carbon savings in successive years in an equitable way.

Petajoule (PJ): a joule x 10^{15}.

Petrol: A mixture of the lighter liquid hydrocarbons, used chiefly as a fuel for internal-combustion engines. It is produced by the fractional distillation of petroleum; by condensation or adsorption from natural gas; by thermal or catalytic decomposition of petroleum or its fractions; by the hydrogenation of producer gas or coal; or by the polymerisation of hydrocarbons of lower molecular weight.

Photovoltaic (PV): The phenomenon of converting light to electric power. Photo = light, Volt = electricity.

Photovoltaics: a photovoltaic solar cell converts light directly into electricity. Light striking the front of a solar cell produces a voltage and current – it has no moving parts. A group of interconnected cells creates a PV panel and PV panels, in turn, can be connected in series or parallel to create a solar array and any voltage-current combination required.

Photovoltaic array in solar power systems: this includes power conditioning unit such as a battery charge controller, inverter and other equipment.

Plant Margin: Plant margin is the spare capacity that electricity generating stations have at any given time. In 2001 the installed plant margin was 27%. By 2003 it was only 20% and it is still falling. Once the peak demand exceeds the installed capacity there is blackout. This is considered increasingly likely in any future cold winter.

Polycyclic aromatic hydrocarbons (PAH): About 100 different organic compounds formed during incomplete combustion. PAHs are widely believed to be carcinogenic. Tobacco smoke is the most common source of PAHs for humans.

Power: The rate at which work is done. It is the product of Voltage times Current, measured in Watts. 1000 Watts = 1 kilowatt. An electric motor requires approximately 1 Kilowatt per Horsepower (after typical efficiency losses). 1 Kilowatt for 1 Hour = 1 kilowatt-hour (kWh).

Premium: A general term to describe the quality of a fuel.

Prime Mover: A prime mover is the drive system for a CHP scheme. The systems that are currently available are all based on engines. There are three commonly used prime movers: gas turbines; reciprocating engines; steam turbines.

Primary energy: Primary energy to deliver a given service is the energy converted when a fuel is burned, for instance, to generate electricity. With the current electricity generation system, primary energy is roughly three times delivered energy: for each unit of electricity delivered to the consumer, two units are lost in generation and transmission.

Pump Station: A mechanical device comprising a pump, flow

controls, non return valves gauges, means of pressurising and filling of thermal loops used in the more sophisticated solar thermal systems.

PV: The common abbreviation for Photovoltaic.

PV Array-Direct: The use of electric power directly from a photovoltaic array, without storage batteries to store or stabilise it. Most solar water pumps work this way, utilising a tank to store water.

PV Array: A group of PV (photovoltaic) modules (also called panels) arranged to produce the voltage and power desired.

PV Cell: The individual photovoltaic device. The most common PV modules are made with 33 to 36 silicon cells each producing 1/2 volt. For further info click here

PV Lighting System: a system that includes at least a PV module, a battery, an inverter and a light. It works best with high efficiency fluorescent, LED, or halogen lamps.

PV light suppliers: there are a number of international photovoltaic suppliers who range from specialist suppliers, to renewable energy equipment suppliers, to suppliers of leisure and camping equipment.

PV Module: An assembly of PV cells framed into a weatherproof unit.

Pyrolysis: The production of gaseous fuels by heating hot materials containing organic matter in the absence of air.

Quad: quadrillion British thermal units.

Reciprocating Engine: One which produces the mechanical power by the up and down or forwards and backwards ("reciprocating") movement of piston within a cylinder, reciprocating engines should be distinguished from rotating machines such as turbines.

Refrigerator: An appliance that cools an area, generally for food, beverage or vaccine storage.

Reformer: A device for processing a fuel such as methane (CH4) into hydrogen (H2) for use in a fuel cell.

Renewable energy: Energy flows that occur naturally and repeatedly in the environment. This includes solar power, wind, wave and tidal power and hydroelectricity. Solid renewable energy sources include energy crops and other biomass; gaseous renewables come from landfill and sewage waste.

Renewable Energy Foundation: A not for profit Organisation which promotes renewable energy whilst safeguarding landscapes from industrialisation, emphasising the need for an effective energy policy that is balanced and eco friendly.

Renewables Obligation: The obligation placed on electricity suppliers to deliver a stated proportion of their electricity from eligible renewable sources.

Renewables Obligation Certificate (ROC): Eligible renewable generators receive ROCs for each MWh of electricity generated. These certificates can be sold to suppliers. In order to fulfil their obligation, suppliers can present enough certificates to cover the required percentage of their output, or pay a 'buyout' price per MWh for any shortfall. All proceeds from buyout payments are recycled to suppliers in proportion to the number of ROCs they present.

Renewable energy: energy from sources such as the sun, wind, water, waves, tides, that are renewable, that is, that can be renewed within a very short period of time. Some definitions include biomass, which requires much longer periods to be renewed.

Resistance: the property of a conductor (i.e. a wire or appliance) which opposes the flow of current through it and converts electrical energy into heat. Resistance has the symbol R, and is measured in ohms.

RET: Renewable energy technologies. Technologies that utilise renewable energy resources.

Royal Commission on Environmental Pollution (RCEP): an independent standing body established in 1970 to advise the Queen, the Government, Parliament and the public on environmental issues.

Salient Pole Generator: A type of electricity generator. As frequency depends on speed multiplied by the number of pole pairs, lower speeds require more poles than can be accommodated within the rotor periphery.

Sankey Diagram: A diagram demonstrating graphically and in true proportion the energy flows in a system, stating with the energy sources at the left and showing losses, heat exchange loops etc. to the degree desired.

Second: The second is the basic unit of time. It is the length of time taken for 9192631770 periods of vibration of the caesium-133 atom to occur.

Semi-conductor: A material having a resistivity, which is high, but not high enough to be classified as an insulator; a solid which is an electrical non-conductor in its pure state or at low temperatures and becomes a conductor when impure or at higher temperatures.

Shell Type Boiler: A cylindrical steam, hot water or thermal oil boiler, usually horizontal but may be vertical. The shell contains water or oil that is heated by the burner flame and combustion products in a chamber and tube or annular flue-ways inside the shell.

Shaft Efficiency: That percentage of its initial energy supply that a prime mover delivers as mechanical energy at its output shaft.

Shell and Tube Heat Exchanger: A unit having a bundle of tubes contained in a cylindrical shell. One fluid flows through the tubes, the other through the shell.

SI: The International System of Units abbreviated from the French *Le Systeme Internationale d'Unites.*

Silicon: A semi-conductor material commonly used to make photovoltaic cells.

Sink: Any activity, mechanism or process which removes a greenhouse gas, an aerosol or a precursor of a greenhouse gas or aerosol from the atmosphere.

Solar cell module: Groups of encapsulated solar cells framed in glass or plastic units, usually the smallest unit of solar electric equipment available to the consumer.

Solar cell: A specially-made semiconductor material (i.e. silicon) which converts light energy into electric energy.

Solar collector: Another name for a thermal solar panel, so named because it collects light and converts it to heat.

Solar electricity: Electricity that is generated by the sun's rays. The most common form of solar electricity in use today is from photovoltaics. However, solar electricity is being produced from solar thermal arrays (typically mirrors focusing the sun's heat on turbines) in several parts of the world.

Solar energy: Energy that is received from the sun. Strictly speaking the term can be used to include wind energy, because wind is derived from the sun, but traditionally it refers only to light energy, mainly solar thermal and PV.

Solar hot water: see solar thermal; solar heated water can be used for domestic industrial leisure or agricultural operations.

Solar module: A collection of solar thermal panels plumbed to generate heat or hot water or a collection of PV cells wired to generate electricity from the sun, connected to a system, or arranged into an array.

Solar radiation: The amount of light energy being received at any given time; also see insolation.

Solar thermal: A system for using solar radiation to heat water, typically in a roof mounted panels connected with pipes to a storage tank. This is the most mature of all renewable energy technologies, also the use of light to create heat.

Solar tracker: A mounting rack for a PV array that automatically tilts to follow the daily path of the sun through the sky. A "tracking array" will produce more energy through the course of the day, than a "fixed array" (non-tracking) particularly during the long days of summer; see tracking.

SOx: A generic term for oxides of sulphur produced by the combustion of sulphur in the fuel, and considered as deleterious emissions. Their presence in flue gases can restrict thermal efficiency, because if the flue gas temperature is reduced below specific levels, highly corrosive sulphurous and sulphuric acids are deposited on heat exchange surfaces.

Spark-ignition: A reciprocating engine that utilises an electrical spark to ignite the compressed air/fuel mixture in the cylinders.

Specific gravity: The ratio of the weight of a solution to an equal volume of water as a specific temperature.

Spectroradiometer: Device for measuring the spectral irradiance (the flux per unit of wavelength: W/m2/micrometer).

Square metre: One metre by one metre on a flat surface (1 m^2).

Stand-alone system: any system that is not connected to a grid or distribution system. The term is most often used for systems not connected to the electricity or natural gas grids (see stand-alone solar electric system).

Steady state economy: a theory of economics which proposes that economic systems should be sustainable with stabilised population and consumption; it also suggests that growth has great disadvantages.

Storage: Any system by which energy is stored. Storage generally includes batteries, although more advanced storage systems can include heat pumps or water reservoirs.

Standard Assessment Procedure (SAP): The SAP is the UK Government's recommended system for energy rating of dwellings. It can be used to calculate the likely energy costs of a building and carbon output.

Stirling engine: An external combustion engine used for generating electricity. Heat moves a piston inside the device, and the moving piston can be used to power an electrical generator. Early designs of micro-CHP units use Stirling engines.

Superheated Steam: Steam whose temperature has been raised above the saturation temperature. This improves its power generating capacity when used in a steam turbine.

Supplementary Firing: The firing of additional fuel in the CHP heat recovery unit, utilising the hot oxygen present as excess air in the prime mover exhaust gases.

Synchronism: The conditions under which generator frequency and voltage levels match those of the public supply.

System: In the case of thermal solar a system refers to the complete components required t produce energy. In the case of electricity, a system means the integrated, connected range of electricity supply elements and the means of transmitting and distributing electricity.

System voltage: the voltage at which the charge controller, lamps and appliances in a system operate, and at-which the module (s) and battery are configured.

Tce: Tonnes of coal equivalent.

Therm: A measurement of energy. There are two therms. The European therm is equivalent to 1.05506 joule and the American therm, which is equivalent to 1.054804 joule.

Thermo chemical Btu: This is based on the thermo chemical calorie which equals 4.184 joule.

Thermal Conductivity: This is ability of a material to conduct heat and is measured in Watts per cubic metre. Copper has a thermal conductivity of 385, Aluminium 211, Steel 47.6 and typical concrete is 1.73. At 20degrees Celsius water has a thermal conductivity of .596 whereas air, at the same temperature has a thermal conductivity of .026.

Thermal diffusivity: This is the ability of a substance to diffuse heat or energy over a given period of time. Air diffuses much more heat than water and as such is unsuitable for thermal solar systems where the heat should be concentrated in the places needed.

Thermal expansion: Warming of the ocean leads to an expansion of the ocean volume and hence an increase in sea level. As the volume of water increases with warming so its density decreases.

Thermal Oil: A mineral oil used as a heat-carrying medium in preference to water or steam. Its major advantage is that high temperatures (up to 300°C) are feasible at pressures far lower than would be needed for steam.

Thermal Ring Main: A system of pipes taking heat, typically in the form of hot water, from an energy production centre to any number of homes, or other end-users and then returning the cooled water to the energy production centre. A thermal ring main is sometimes called a heat network, used in District Heating Systems. Efficient thermal ring mains depend upon the temperature of the return being significantly lower than the input temperature.

Tracking: The practise of changing the position (i.e. angle) of the array at various times during the day so that is faces the sun and so collects a larger amount of light.

Transformer: An electrical device that steps up voltage and steps down current proportionally (or vice-versa). Transformers work with AC only.

Useful energy: The net energy provided to a dwelling for space or water heating, from a heat source (that is to say the actually delivered multiplied by the boiler efficiency).

Utility grid: A utility grid is usually a commercial electric power distribution system that takes electricity from a generator and transmits it over a certain distance, and then takes the electricity down to the consumer through a distribution system.

U-value: The U value of a building element is an expression of the rate of energy flow (Watts) for a given surface area (m2) and a given temperature difference between indoors and outdoors, conventionally expressed on the Kelvin scale (K), but practically measured in degrees Celsius. U values are measured in W/m2K. The lower the U-value is, the better the thermal insulation is.

VAC: Volts alternating current.

Vacuum: a region in which the gas pressure is considerably lower than atmospheric pressure.

Vacuum panel: A thermal solar flat plat panel in which the vacuum is used for insulating against heat loss.

Vacuum Tube: a type of thermal solar collector with the absorber arranged inside an evacuated tube to prevent heat losses.

Vacuum insulation panel (VIP): A VIP consists of a special panel enclosed in an air-tight envelope, to which a vacuum is applied and which is used for insulation.

Volt (V): A unit of measurement of the force given to electrons in an electric circuit.

Voltage: Voltage is the measurement of electrical potential.

Voltage drop: Voltage drop is the loss of voltage (electrical pressure) caused by the resistance in wire and electrical devices. Proper wire sizing will minimise voltage drop, particularly over long distances. Voltage drop is determined by 4 factors: wire size, current (amps), voltage, and length of wire. It is determined by a consulting wire sizing chart or formula available in various reference tests. It is expressed as a percentage.

Voltage, nominal: Nominal voltage is a way of naming a range of voltage to a standard. Example: A "12 Volt Nominal" system may operate in the range of 11 to 15 Volts. We call it "12 Volts" for simplicity.

Voltage, open circuit: The voltage of a PV module or array with no load (when it is disconnected). A "12 Volt Nominal" PV module will produce about 20 Volts open circuit. Abbreviation: Voc.

Voltage, peak power point: The voltage at which a photovoltaic module or array transfers the greatest amount of power (watts). A "12 Volt Nominal" PV module will typically have a peak power voltage of around 17 volts. A PV array-direct solar pump should reach this voltage in full sun conditions. In a higher voltage array, it will be a multiple of this voltage. Abbreviation: Vpp.

Warm Deal: A scheme for the provision of energy efficiency improvements to households whose members are receiving benefits; similar to Warm Front but designed for Scotland.

Warm Front: A scheme for the reduction of fuel poverty in England in vulnerable households by improving energy efficiency. It is aimed at households with children, the over-60s and the disabled or long-term sick It also aims to reduce the incidence of fuel debt within the target group, improve comfort levels and prevent cold-related illness. Annual expenditure is approximately £150m.

Warm Homes: A scheme for the provision of energy efficiency improvements to households whose members are receiving state benefits, similar to Warm Front but covering Northern Ireland.

Water Tube Boiler: this operates in the opposite way to a shell type boiler. A box acts as a combustion chamber.

Watt (W): The internationally accepted measurement of power. One thousand watts are a kilowatt, and a million watts are a megawatt. A Watt is the power used when one Joule of energy is used every second (i.e., 1 Watt = Joule/second; 1 Joule = 1 Watt second; 1 Watt hour/Wh = 3.6 thousand Joules/kJ; 1 kWh = 3.6 MegaJoules/MJ).

Watt hour (Wh): A common energy measure arrived at by multiplying the power times the hours of use (1 Watt hour/Wh = 3.6 kilo Joules/kJ; 1 kWh = 3.6 MegaJoules/MJ). Grid power is ordinarily sold and measured in kilowatt hours.

Wind power: Power generated by the movement of the earth's rotation, and the temperature variations around the globe. Wind power is usually mechanically converted into electricity by generators, or used to power vessels or mills.

Wind turbines: Wind mills are driven by the force of the wind to generate power, either mechanically (as for pumping) or electrically (as in wind electricity generators).

W/m2K: See U value.

Index